e-Humor!

Version 1.0

By

Joseph J. Zajac III

This book is a work of fiction. Places, events, and situations in this story are purely fictional. Any resemblance to actual persons, living or dead, is coincidental.

ISBN: 1-4033-1307-5 (e-book)
ISBN: 1-4033-1308-3 (Paperback)

This book is printed on acid free paper.

1stBooks - rev. 10/23/02

Disclaimer

Material contained in this book was obtained in the public domain over the Internet and to the best of my knowledge is not copyrighted. This is a book on humor so take no offense to the contents. If you cannot take a joke…you know the rest of the line.

Dedication

This collection of humor is dedicated to the following:

- ➢ To my parents, Joe and Helen, who taught me that there is nothing wrong in working hard;
- ➢ To my wife, Pamela, who takes care of me;
- ➢ To my dog, Athena, who always gives unconditional love;
- ➢ To my sister, Barbara Jean, who took care of me in my younger years;
- ➢ To my brother, Robert, who taught me never to turn your back on someone holding a BB rifle;
- ➢ To my friends, Walter, Mark, Dan, Jim and Jon, who inspired me to put the real value of the Internet into a book;
- ➢ To John and Marty, who are two of the most inspirational people I know; and
- ➢ To those individuals who keep America safe from our enemies, both foreign and domestic.

Table of Contents

Introduction

Thank you for purchasing my book!

Based upon my two decades of experience with the world of "high technology," I have synthesized all that knowledge into the following two sections:

Undeniable Truths in Our Technological World # 1

> ➤ Al Gore did not invent the Internet.
> ➤ An Indian programmer always shakes his head to mean "Yes" and nods his head to mean "No."
> ➤ Anything called a "BETA RELEASE" is Greek for RUN AWAY AS FAST AS YOU CAN!
> ➤ Anytime you are told something is "State of the Art," RUN AWAY AS FAST AS YOU CAN!
> ➤ Anytime you are told the software "Is easily implementable," RUN AWAY AS FAST AS YOU CAN!
> ➤ Anytime you are told the software is "Robust" or "Sexy," RUN AWAY AS FAST AS YOU CAN!
> ➤ Anytime you hear about a "Vision," RUN AWAY AS FAST AS YOU CAN!
> ➤ Ask ten consultants what the "e" in e-"you fill in the blank" means and you will receive ten different answers.
> ➤ Bill Gates of Microsoft WILL ALWAYS BE RICHER than Oracle's Larry Ellison, and rightly so!
> ➤ Computer equipment salespeople lie for a living.
> ➤ Computer software salespeople lie for a living.
> ➤ Computer telecommunications salespeople lie for living.
> ➤ Consultants are only two weeks ahead of your learning curve on anything.
> ➤ Does anyone really care what HTTP means?
> ➤ e-Banking really means customer self service.
> ➤ e-Business really means self service.
> ➤ e-Commerce really means customer self service.
> ➤ e-Customer really means customer self service.
> ➤ e-Education really means customer self service.
> ➤ e-Humor is the only good use of the Internet.
> ➤ e-IEIO really means customer self service.
> ➤ e-Manufacturing really means customer self service.
> ➤ e-MyAss still means customer self service.
> ➤ e-Procurement really means customer self service.

Undeniable Truths in Our Technological World # 2

➢ Free software really means "We do not have people to conduct quality testing of our product so please tell us when you find any bugs."

➢ If you get "personal" with your "Personal Computer," please see a therapist.

➢ It is beneficial to learn some words in Hindi as the Indians are leading the world in offshore software development.

➢ Most technology start up companies will fail.

➢ Most technology start up companies hire executives who failed at other start up companies.

➢ Never purchase any software at Release 1.0., Release 2.0 or Release 3.0.

➢ Never purchase any software until the first "patch" (bug fixes) release comes out.

➢ Never allow your business to be a "test site" or a "beta site" for new hardware or software.

➢ No matter how fast or expensive the computer system, when Garbage Goes In, Garbage Comes Out.

➢ People at the other end of the telephone on the "Help line" really do not know more than you do.

➢ People still make the same mistakes today as they ten years ago only now the mistakes are processed faster.

➢ Software and hardware salespeople started their careers selling used cars.

➢ Software and hardware salespeople still seem better suited to selling used cars.

➢ Software companies purposely make new software more complex so that you have to upgrade your computer or purchase a new computer just to run their new software.

➢ There are no "standards" for any new technology, just get used to the new problems caused by the new and improved technology.

➢ Time to upgrade just so you can use a new printer or scanner with a USB connection.

➢ User Committees really mean the company producing the software has no clue on how to improve their product.

➢ User manuals are seldom user friendly.

➢ User manuals seem to be written overseas by people who have English as their third language.

➢ Whatever you buy, it will be obsolete in three months and the newer version will be cheaper.

➢ When in doubt, REBOOT! REBOOT! REBOOT!

➢ www spelled backwards is still www.

Science and Technology

Why computers must be female

1. As soon as you make a commitment to one, you find yourself spending half your paycheck on accessories for it.
2. Even your smallest mistakes are immediately committed to memory for future reference.
3. No one but their creator understands their internal logic.
4. The message, "Bad command or filename," is about as informative as "If you don't know why I'm mad at you, then I'm certainly not going to tell you."
5. The native language used to communicate with other computers is incomprehensible to everyone else.

Smoke

My neighbor works in the operations department in the central office of a large bank. Employees in the field call him when they have problems with their computers. One night he got a call from a woman in one of the branch banks who had this question:

"I've got smoke coming from the back of my terminal. Do you guys have a fire downtown?"

Technology Lies

➢ As long as you remember to 'SAVE' your input, you will never lose any files.
➢ Of course we can modify it.
➢ The new machine's on order.
➢ The program's fully tested and bug free.
➢ The programs work as designed.
➢ We cannot duplicate your problem on our system.
➢ We throughly test our hardware.
➢ We throughly test our software.
➢ We throughly tested our software for your hardware configuration.
➢ We are working on the fixes as fast as we can.
➢ We are working on the documentation.

Microsoft

How many Microsoft software engineers does it take to screw in a light bulb?

None...they just upgrade the standard to "darkness," and hire 15 more customer service employees to inform the customers.

Office Lingo

404	Someone who is clueless, from the World Wide Web error message "404 Not Found", meaning the requested document couldn't be located — Don't bother asking him, he's 404.
Blamestorming	Sitting around in a group discussing why a deadline was missed or a project failed and who was responsible
Body Nazis	Hard-core exercise and weight-lifting fanatics who look down on anyone who doesn't work out obsessively
Chainsaw consultant	An outside expert brought in to reduce the employee headcount, leaving the top brass with clean hands
Cube farm	An office filled with cubicles
Ego surfing	Scanning the Net, databases, print media, and so on, looking for references to one's own name
Elvis year	The peak year of something's popularity — Barney the dinosaur's Elvis year was 1993.
Idea hamsters	People who always seem to have their idea generators running
Mouse potato	The on-line generation's answer to the couch potato
Ohnosecond	That minuscule fraction of time in which you realize you've just made a big mistake
Prairie dogging	Something loud happens in a cube farm, and people's heads pop up over the walls to see what's going on
SITCOM	Stands for Single Income, Two Children, Oppressive Mortgage

4

Stress puppy	A person who thrives on being stressed-out and whiny
Tourists	Those who take training classes just to take a vacation from their jobs — "We had three serious students in the class; the rest were tourists."
Uninstalled	Euphemism for being fired
Xerox subsidy	Euphemism for swiping free photocopies from a workplace

Tibet

Two archeologists, exploring a remote mountain in Tibet came across a huge granite statue which resembled a sitting man. It stood almost 400 foot tall, and its bodily details were accurate down to the fingernails and teeth.

"It looks real enough to talk", said one archaeologist.

"Lets try," said the other as he turned to the statue and asked its name. No answer.

"How old are you?" No answer.

Finally. one shouts out, "What is the square root of 64?!"

Suddenly, the mountain shakes as the giant statue rose to its feet and put its hand on its chin. After about ten seconds, the statue answers in a roaring voice, "Eight."

"Of course," says the first scientist, "it only stands to reason."

Drug dealers vs. Software developers

Drug Dealers	Software Developers
Refer to their clients as "users."	Refer to their clients as "users."
"The first one's free!"	"Download a free trial version..."
Have important South-East Asian connections (to help move the stuff).	Have important South-East Asian connections (to help debug the code).
Strange jargon: "Stick," "Rock," "Dime bag," "E"	Strange jargon: "SCSI," "RTFM," "Java," "ISDN."

Joseph J. Zajac III

Realize that there's tons of cash in the 14- to 25-year-old market.	Realize that there's tons of cash in the 14- to 25-year-old market.
Job is assisted by the industry's producing newer, more potent mixes.	Job is assisted by industry's producing newer, faster machines.
Often seen in the company of pimps and hustlers.	Often seen in the company of marketing people and venture capitalists.
Their product causes unhealthy addictions.	DOOM. Quake. SimCity. Duke Nukem 3D.
Do your job well, and you can sleep with sexy movie stars who depend on you.	Damn! Damn! Damn!

IBM vs. Sun

Three IBM salespeople and three SUN salespeople are traveling by train to a conference. At the station, the three IBM folks each buy tickets and watch as the three SUN people buy only a single ticket. "How are three people going to travel on only one ticket?" asks an IBMer. "Watch and you'll see." answers an SUN person. They all board the train.

The IBM'ers take their respective seats but all three SUN people cram into a rest room and close the door behind them.

Shortly after the train has departed, the conductor comes around collecting tickets. He knocks on the restroom door and says, "Ticket, please." The door opens just a crack and a single arm emerges with a ticket in hand. The conductor takes it and moves on. The IBM's saw this and agreed it was quite a clever idea.

After the conference, the IBM'ers decide to copy the SUN folks on the return trip and save some money. When they get to the station, they buy a single ticket for the return trip. To their astonishment, the SUN people don't buy a ticket at all. "How are you going to travel without a ticket?" says one perplexed IBM'er. "Watch and you'll see," answers an SUN person.

When they board the train the three SUN'ers cram into a restroom and the three IBM'ers cram into another one nearby. The train departs. Shortly afterward, one of the SUN'ers leaves his restroom and walks over to the restroom where the IBM'ers are hiding. He knocks on the door and says, "Ticket, please."

Girlfriend Upgrade

Last year I upgraded my GirlFriend 5.0 to GirlFriend 5.1, which installed itself as Fiancée 1.0. Recently, I upgraded Fiancée 1.0 to Wife 1.0 and its a real memory hog. It has taken up all my space, and Wife 1.0 must be running before I can do anything.

Wife 1.0 is also spawning Child Processes which are further consuming system resources. Other applications, such as PokerNight 10.3, BeerBash2.5 and PubNight 7.0, are no longer able to run in the system at all.

Additional Plug-ins were automatically installed, such as Mother-in-Law 55.8 - for which there is no uninstall feature. None of these behaviors were discussed in the documentation, although other users have reported similar problems.

Because of this, some users I know have decided to avoid the headaches associated with these upgrades, and simply move from GirlFriend 5.0 to GirlFriend 6.0. Unfortunately this is not without peril as all traces of GirlFriend 5.0 must be removed from the system before attempting installation of 6.0.

Even then, GirlFriend 6.0 will repeatedly run system checks (usually in the background, and often late at night when the system is asleep) to find evidence of previous versions. To cap it off, GirlFriend 6.0 apparently has a Nag feature reminding the user about advantages of upgrading to Wife 1.0.

However, I do like some of the planned features to be released in the upcoming GirlFriend 6.1 release:

> ➢ A "Don't remind me again" button
> ➢ A "Minimize" button
> ➢ A Shutdown feature
> ➢ An install shield feature so that GirlFriend can be completely uninstalled if necessary (so you don't lose cache and other objects).

Unfortunately, since I already upgraded to Wife 1.0, I don't think I will be able to take advantage of any of these new features unless you decide to include them in the next Mistress release. But, of course, there is a whole raft of problems associated with the use of Mistress 1.0 and Wife 1.0 on the same system. Most notable are system conflicts and continual disk thrashing, which starts shortly after Wife 1.0 detects Mistress 1.0. Interestingly enough, all versions of PersonalLawyer still work fine.

Finally, Wife 1.0 apparently deletes all MSMoney files before uninstalling itself; following that, Mistress 1.1 will refuse to install, claiming insufficient resources.

I personally find all these new tools and conflicts to be too confusing and time consuming. I'm sticking with Dog 1.0b3. It slobbers and chews up the paper, but in all, these bugs are tolerable. It is simple to operate and we get along fine.

Project Management

Three men: a project manager, a software engineer, and a hardware engineer are in Ft. Lauderdale for a two-week period helping out on a project.

About midweek they decide to walk up and down the beach during their lunch hour. Halfway up the beach, they stumbled upon a lamp. As they rub the lamp a genie appears and says "Normally I would grant you 3 wishes, but since there are 3 of you, I will grant you each one wish."

The hardware engineer went first. "I would like to spend the rest of my life living in a huge house in St. Thomas, with no money worries and surrounded by beautiful women who worship me." The genie granted him his wish and sent him on off to St. Thomas.

The software engineer went next. "I would like to spend the rest of my life living on a huge yacht cruising the Mediterranean, with no money worries and surrounded by beautiful women who worship me." The genie granted him his wish and sent him off to the Mediterranean.

Last, but not least, it was the project manager's turn. "And what would your wish be?" asked the genie.

"I want them both back after lunch" replied the project manager.

Computer Viruses

Be on the lookout for these viruses:

Adam And Eve Virus
Takes a couple bytes out of your Apple.

Airline Luggage Virus
You're in Dallas, but your data is in Singapore.

Bill Clinton Virus
Causes your PC to behave unpredictably, working as expected one moment, then suddenly doing the exact opposite the next moment.

Congressional Virus
The computer locks up, and the screen splits in half with the same message appearing on each side of the screen. The message says that the blame for the gridlock is caused by the other side.

Democrat Virus
Doesn't allow you to delete inefficient programs or wasted disc space - if you try, it accuses you of being a "mean-spirited extremist."

Elvis Virus
> Your computer gets fat, slow, and lazy, then self destructs, only to resurface at shopping malls and service stations across rural America.

Federal Bureaucrat Virus
> Divides your hard disk into hundreds of little units, each of which does practically nothing, but all of which claim to be the most important part of your computer.

Freudian Virus
> Your computer becomes obsessed with marrying its own motherboard.

Gallup Virus
> Sixty percent of the PC's infected will lose 30 percent of their data 14 percent of the time (plus or minus a 3.5 percent margin of error).

George Bush Virus
> It starts by boldly stating, "Read my docs...no new files!" on the screen. It proceeds to fill up all the free space on your hard drive with new files, then blames it on the congressional virus.

Health Care Virus
> Tests your system for a day, finds nothing wrong, and sends you a bill for $4,500.

Jack Kevorkian Virus
> Assists your CPU in destroying itself.

Jimmy Hoffa Virus
> Your programs can never be found again.

Jocelyn Elders Virus
> Teaches your computer to turn itself on.

Kevorkian Virus
> Helps your computer shut down as an act of mercy.

LAPD Virus
> Attempts to stop your CPU. If your CPU resists, it is pummeled into hamburger.

Lyle And Eric Menendez Virus
> Wipes out your motherboard, claiming it was done in self-defense.

Militia Virus
Wipes out your operating system claiming it has no right to control your PC.

National Education Assoc. (NEA) Virus
Although cleverly disguised as educational software intended to improve your system, in reality it "dumbs down" your 486DX into an 8086.

National Organization Of Women (NOW) Virus
Forces your PC to recognize its female connections as male connections.

Nike Virus
Just does it.

Ollie North Virus
Causes your printer to become a paper shredder.

Politically Correct Virus
Rephrases the "Abort, Retry, Fail" prompt as "Choice, Retry, Success-Impaired."

Pro-Choice Virus
Although it presents the standard "Abort, Retry, Fail" prompt, it pressures you to choose "Abort", telling you the process being terminated is just "a blob of bits" which has no value.

PBS Virus
Your programs stop every few minutes to ask for money.

Republican Virus
Sells off your system resources to the highest bidder.

Ross Perot Virus
This erratic virus doesn't do much of anything, except surfacing occasionally to threaten to disrupt your system.

Sears Virus
Your data won't appear unless you buy new cables, power supply, and a set of shocks.

Star Trek Virus
Invades your system in places where no virus has gone before.

Texas Virus
Makes sure that it's bigger than any other file.

Mistress

An artist, a lawyer, and a computer scientist are discussing the merits of a mistress.

The artist tells of the passion, the thrill that comes with the risk of being discovered.

The lawyer warns of the difficulties. It can lead to guilt, divorce, bankruptcy. Not worth it. Too many problems.

The computer scientist says "It's the best thing that's ever happened to me. My wife thinks I'm with my mistress and I can spend all night on the computer!"

Keep it Simple Stupid

During the heat of the space race in the 1960's, the U.S. National Aeronautics and Space Administration decided it needed a ball point pen to write in the zero gravity confines of its space capsules. After considerable research and development, the Astronaut Pen was developed at a cost of about $241 million. The pen worked and enjoyed some modest success as a novelty item back here on Earth.

The Soviet Union, faced with the same problem, used a pencil!!!

Three Consultants

Three IT Consultants were in the urinal performing their morning constitutional.

The first Consultant finishes and walks over to the sink to wash his hands. He then proceeds to dry his hands very carefully. He uses paper towel after paper towel and ensures that every single spot of water on his hands are dried. Turning to the other two Consultants, he says, "At Blunder Consulting, we are trained to be extremely thorough."

The second Consultant finishes his task at the urinal and he proceeds to wash his hands. He uses a single paper towel and makes sure that he dries every drop of water from his hands using every available portion of the paper towel. He turns and says, "At IBS, not only are we trained to be extremely thorough but we are also trained to be extremely efficient."

The third Consultant finishes and walks straight for the door. "At Apex, we learn not to piss on our hands."

Why Being A Prostitute Is Better Than Being A Consultant

> ➤ Close client interaction at all times
> ➤ Continual feedback - every two hours or so
> ➤ Either way you are screwing clients
> ➤ Finally a way to fit exercise into a tight schedule

> Hotel, etc. expenses are directly billed
> No dress code
> Not tied down working with a team (unless you want to be)
> You actually receive that high hourly rate clients are paying for you
> You are working nights anyway
> You get to choose your clients

Q and A

Q: How many McKinsey consultants does it take to change a light bulb?
A: How many can you afford?

Q: If you see a consultant on a bicycle, why should you never swerve to hit him?
A: It might be your bicycle.

Q: Why are consultants like nuclear weapons?
A1: If one side has one, the other side has to get one.
A2: Once launched, they cannot be recalled.
A3: When they land, they screw up everything forever.

Q: Why did the Post Office just recalled their latest stamps?
A: They had pictures of consultants on them ... and people couldn't figure out which side to spit on.

Q: Why do consulting companies prohibit sex between consultants and their clients?
A: To prevent clients from being billed twice for essentially the same service.

Q: How many consultants does it take to change a light bulb?
A: We don't know. They never get past the feasibility study.

Things a Consultant Should Never Tell a Client

> Hey, I just realized that I was in junior high when you started working here.
> I could just tell you the answer, but we're committed to a three month project.
> I like this office space. I'll have them put me in here when you're gone.
> My rental car looks nicer than that junker you're driving.
> Of course it's right; the spreadsheet says so.
> So what do you need me to tell you?
> Sure it'll work; I learned it in business school.

> That was my first guess as well, but then I really thought about it.
> What are you, stupid?
> You should see the hotel I'm staying at.

Consulting Rates

A man walked into a consultant's office and inquired about the rates for a study.

"Well, we usually structure the project up front, and charge $50K for three questions," replied the consultant.

"Isn't that awfully steep?" asked the man.

"Yes," the consultant replied, "and what was your third question?"

Things You Should Never Say at a Consulting Interview

> Call it what you want, it still means firing people.
> Do you cover rental cars for collision?
> Do you pay overtime?
> I am afraid of flying.
> I think three letter acronyms are for people too stupid to remember whole phrases.
> I am a t-shirt and jeans kind of person.
> I am useless without ten hours of sleep a night.
> Stanford taught me that working in teams is great for slackers.
> There are lies, damn lies, and statistics.
> Two words: family first.

Ways To Know You're Dating/Married To A Consultant

> Cannot be trusted with the car-too accustomed to beating up rentals.
> Celebrates anniversary by conducting a performance review.
> Congratulates your parents for successful value creation.
> Ends any argument by saying, "let's talk about this off-line."
> Referred to the first month of your relationship as a "diagnostic period."
> Refers to lovemaking as a "win-win."
> Takes a half-day at the office because, "Sunday is your day."
> Talks to the waiter about process flow when dinner arrives late.
> Tries to call room-service from the bedroom.
> Valentine's Day card has bullet points.

Ways to Know You've Got the Consulting Bug

> ➤ A two-page story in Business Week is all it takes to make you an expert.
> ➤ Always-hyphenating-words-that-don't-need-to-be-hyphenated.
> ➤ Can fit the thematic undercurrents of War and Peace into a two-by-two matrix.
> ➤ Cannot stop using words that don't exist.
> ➤ Constant urge to give advice on subjects you know nothing about.
> ➤ Firmly believe that an objective viewpoint means more than any real work experience.
> ➤ Keep seeing bullet points everywhere.
> ➤ Tired of having a social life beyond work.
> ➤ Use so much jargon in conversation, friends think you're speaking a foreign language.
> ➤ Worried that he who dies with the most frequent-flyer miles wins.

Things You Will Never Hear from a Consultant

> ➤ Actually, the only difference is that we charge more than they do.
> ➤ Bet you I can go a week without saying "synergy" or "value-added."
> ➤ Everything looks okay to me.
> ➤ How about paying us based on the success of the project?
> ➤ I cannot take the credit. It was Ed in your marketing department.
> ➤ I do not know enough to speak intelligently about that.
> ➤ Implementation? I only care about writing long reports.
> ➤ The problem is, you have too much work for too few people.
> ➤ This whole strategy is based on a Harvard business case I read.
> ➤ You are right; we're billing way too much for this.

Bill Gates

Bill Gates dies and finds himself being sized up by God.

"Well, Bill, I'm really confused on this call. I'm not sure whether to send you to Heaven or Hell. After all, you enormously helped society by putting a computer in almost every home in America, yet you also created that ghastly Windows '95. I'm going to do something I've never done before. I'm going to let you decide where you want to go."

Bill replied, "Well, what's the difference between the two?" God said, "I'm willing to let you visit both places briefly, to see if it will help your decision." "Fine, but where should I go first?" Bill asked. "I'll leave that up to you," God replied. "Okay then," said Bill, "let's try Hell first."

14

So Bill went to Hell. It was a beautiful, clean, sandy beach with clear waters and lots of beautiful women running around, playing in the water, laughing and frolicking about. The sun was shining; the temperature perfect. He was very pleased. "This is great," he told God. "If this is hell, I really want to see heaven."

"Fine," said God, and off they went. Heaven was a place high in the clouds, with angels drifting about, playing harps and singing. It was nice, but not as enticing as Hell. Bill thought for a quick minute, and rendered his decision. "Hmmm. I think I'd prefer Hell," he told God.

"Fine," replied God, "as you desire." So Bill Gates went to Hell.

Two weeks later, God decided to check on the late billionaire to see how he was doing in Hell. When he got there, he found Bill, shackled to a wall screaming amongst hot flames in dark caves, being burned and tortured by demons, with no one to help him out of his dilemma no matter how loud he screamed.

"How's everything going?" He asked Bill.

Bill responded with his voice filled with anguish and tormented disappointment. "This is awful. This is nothing like the Hell I visited two weeks ago. I can't believe this is happening. What happened to that other place, with the beaches and the beautiful women playing in the water????"

"Oh," God said, "that was Hell 3.1. This is Hell 95."

New Software

A great new software announcement!!!! This memo is to announce the development of a new software system.

We are currently building a data center that will contain all company data that is Year 2000 compliant. The program is referred to as the "Millennia Year Application Software System" (MYASS).

Next Monday at 9:00 there will be a meeting in which I will show MYASS to everyone. We will continue to hold demonstrations throughout the month so that all employees will have an opportunity to get a good look at MYASS.

As for the status of the implementation of the program, I have not addressed the networking aspects so currently only one person at a time can use MYASS. This restriction will be removed after MYASS expands.

Several people are using the program already and have come to depend on it. Just this morning I walked into a subordinate's office and was not surprised to find that he had his nose buried in MYASS. I've noticed that some of the less technical personnel are somewhat afraid of MYASS.

Just last week, when asked to enter some information into the program, I had a secretary say to me "I'm a little nervous, I've never put anything in MYASS before." I volunteered to help her through her first time and when we were through she admitted that it was relatively painless and she was actually looking forward to doing it again. She went so far as to say that after using SAP and Oracle, she was ready to kiss MYASS.

I know there are concerns over the virus that was found in MYASS upon initial installation, but I am pleased to say the virus has been eliminated and we were able to save MYASS. In the future, however, protection will be required prior to entering MYASS. We planned this database to encompass all information associated with the business.

So as you begin using the program, feel free to put anything you want into MYASS. As MYASS grows larger, we envision a time when it will be commonplace to walk by an office and see a manager hand a paper to an employee and say "Here, stick this in MYASS."

This program has already demonstrated great benefit to the company during recent OSHA and EPA audits. After requesting certain historical data the agency representatives were amazed at how quickly we provided the information. When asked how the numbers could be retrieved so rapidly our Environmental Manager proudly stated "Simple, I just pulled them out of MYASS."

Observations

> - A 3.5 hard is better than a 5.25 floppy
> - A bus station is where the bus stops. A train station is where the train stops. On my desk, I have a workstation.
> - A)bort, R)etry, F)ake it like it's working...
> - A)bort, R)etry, I)gnore, V)alium?
> - A)bort, R)etry, I)nfluence with large hammer
> - Bugs are sons of glitches
> - Computers are useless. They can only give you answers.-Pablo Picasso
> - Do files get embarrassed when they get unzipped?
> - Do I use XXXMODEM to download the real sexy stuff?
> - Don't let the computer bugs bite!
> - Error 13: Illegal brain function. Process terminated.
> - Hardware: The parts of a computer system that can be kicked.
> - How do I set my laser printer on stun?
> - I have not lost my mind; it's backed up on tape somewhere.
> - I modem, but they grew back.
> - It said, "Insert disk #3," but only two will fit!
> - USER ERROR: replace user and press any key to continue.

Working

Best Reasons For Not Coming To Work

1. Constipation has made me a walking time bomb.
2. I am converting my calendar from Julian to Gregorian.
3. I am extremely sensitive to a rise in the interest rates.
4. I am stuck in the blood pressure machine down at the Food Giant.
5. I cannot come in because the deadline is Monday and so far I only have seven different fun things to do with a barrel of snot.
6. I cannot come in to work today because I'll be stalking my previous boss, who fired me for not showing up for work. Ok?
7. I cannot come to work today because the EPA has determined that my house is completely surrounded by wet lands and I have to arrange for helicopter transportation.
8. I have a rare case of 48 hour projectile leprosy, but I know we have that deadline to meet...
9. I just found out that I was switched at birth. Legally, I shouldn't Come in to work knowing my employee records may now contain false information.
10. I prefer to remain an enigma.
11. I refuse to travel to my job in the District until there is a Commuter tax. I insist on paying my fair share.
12. I set half the clocks in my house ahead an hour and the other half back an hour Saturday and spent 18 hours in some kind of space time continuum loop, reliving Sunday (right up until the explosion).
13. I was able to exit the loop only by reversing the polarity of the power source exactly e*log(pi) clocks in the house while simultaneously rapping my dog on the snout with a rolled up Times. Accordingly, I will be in late, or early.
14. If it is all the same to you I won't be coming in to work. The voices told me to clean all the guns today.
15. I'm feeling a little disgruntled this morning. You sure I should come in?
16. I've used up all my sick days...so I'm calling in dead!
17. My mother in law has come back as one of the undead and we must track her to her coffin to drive a stake through her heart and give her eternal peace. One day should do it.
18. My stigmata's acting up.
19. My wife makes more money than I do, so I have to stay at home with our sick son.
20. The dog ate my car keys. We're going to hitchhike to the vet.

21. The psychiatrist said it was an excellent session. He even gave me this jaw restraint so I won't bite things when I am startled.
22. When I got up this morning I took two Ex Lax in addition to my Prozac. I cannot get off the john, but I feel good about it.
23. Yes, I seem to have contracted some attention deficit disorder and, hey, how about them Skins, huh? So, I won't be able to, yes, could I help you? No, no, I'll be sticking with AT&T, but thank you for calling.

Sales and Management

A woman named Julia was in a hot air balloon and realized she was lost. She reduced altitude and spotted a man named Tony below. Julia descended a bit more and shouted, "Excuse me, can you help? I promised a friend I would meet her an hour ago, but I don't know where I am."

Tony below replied, "You are in a hot air balloon hovering approximately 30 feet above the ground. You are between 40 and 41 degrees north latitude and between 59 and 60 degrees west longitude."

"You must be an engineer," said Julia, the balloonist.

"I am," replied Tony, the engineer, "How did you know?"

"Well," answered Julia, "everything you told me is technically correct but I have no idea what to make of your information, and the fact is, I am still lost. Frankly, you've not been much help so far."

Tony below responded, "You must be in Sales and Management."

"I am," replied Julia, "but how did you know?"

"Well," said Tony, "you don't know where you are or where you are going. You have risen to where you are due to a large quantity of hot air. You made a promise which you have no idea how to keep, and you expect people beneath you to solve your problems. The fact is, you are in exactly the same position you were in before we met, but now, somehow, it's my fault."

Jesus

One day this guy walks into a bar and sits down at one end. A group of three union members are sitting at the other end of the bar. One of them notices that it is Jesus Christ, and says "Hey that's Jesus." The others realize this and they decide to be nice and buy him a beer. They call the bartender over and tell him to send him a beer. So the bartender fills up a beer and shoots it on down to Jesus.

Jesus takes the beer and drinks it right down. After he is done, he gets up and walks over to the three guys. He says to them "I really appreciate what you've done for me, I'd like to help you out." He senses that the first guy has a bad elbow, so Jesus touches the fellow on the arm. The guy feels no more pain in his arm, and he gets up, and he's swinging it around and goes running around the bar shaking everyone's hands.

18

Jesus then walks over to the second man, and senses that he has a bad knee. He places his hand on the man's knee. The man stands up and there is no pain. He thinks it great and starts dancing around the bar, having a great time.

Jesus walks up behind the third but senses nothing wrong with the man. So deciding to ask the man what he would like as he reaches for the man. The man jumps back and shouts "Hey! Don't touch me! I'm on Workers Compensation!

Business Lies

> Delivery of your product will occur within 30 days of ordering it.
> If it were up to me, there'd be no assigned parking spaces.
> If you use our product you will have physical relationships with the same kinds of people that you see in our ad.
> If you're not satisfied with our product we will guarantee you a full refund.
> Immediate delivery? ... No problem.
> Money ... it's just a score card.
> Our staff are courteous and considerate.
> People are our greatest resource.
> Staying small is a conscious decision.
> The boss is just one of the guys.
> This product will taste as good as it looks.
> We have an entrepreneurial spirit here.
> We offer repair of your product free of charge with an accredited repairer in your home State.
> We say 'let the marketplace decide'.
> We treat every customer as if they were our most important.
> We try to help you with your problem.
> We're going out to lunch to talk business.
> You can exchange or get full refund on an item that you're not satisfied with.
> You have to twist my arm to get me to go on a business trip.
> You really need our product.

Everything I Need to Know, I Learned in Corporate America

1. Anything worth fighting for is worth fighting dirty for.
2. Anything you do can get you fired; this includes doing nothing.
3. By the time you make ends meet, they move the ends.
4. Everything should be made as simple as possible, but no simpler.
5. Friends may come and go, but enemies accumulate.
6. Happiness is merely the remission of pain.
7. I have seen the truth and it makes no sense.

8. If you can smile when things go wrong, you have someone in mind to blame.
9. If you think there is good in everybody, you haven't met everybody.
10. Indecision is the key to flexibility.
11. Money cannot buy happiness; it can, however, rent it.
12. Never pass a snow plow on the right.
13. Never wrestle a pig. You both get dirty and the pig likes it.
14. No amount of advance planning will ever replace dumb luck.
15. Nostalgia isn't what it used to be.
16. Not one shred of evidence supports the notion that life is serious.
17. One seventh of your life is spent on Monday.
18. Someone who thinks logically is a nice contrast to the real world.
19. Sometimes too much to drink is not enough.
20. The careful application of terror is also a form of communication.
21. The facts, although interesting, are irrelevant.
22. The trouble with life is, you're halfway through it before you realize it's a do-it-yourself thing.
23. There is absolutely no substitute for a genuine lack of preparation.
24. There is always one more imbecile than you counted on.
25. Things are more like they are today than they ever were before.
26. This is as bad as it can get, but don't count on it.
27. You cannot tell which way the train went by looking at the track.
28. Youth and skill are no match for experience and treachery.

Christmas Party

After the annual office Christmas party blowout, John woke up with a pounding headache, cotton-mouthed, and utterly unable to recall the events of the preceding evening. After a trip to the bathroom he was able to make his way downstairs, where his wife put some coffee in front of him

"Louise," he moaned, "tell me what went on last night. Was it as bad as I think?"

"Even worse," she assured him, voice dripping with scorn. "You made a complete ass of yourself, succeeded in antagonizing the entire board of directors, and insulted the president of the company to his face."

"He's an asshole - piss on him."

"You did," Louise informed him. "And he fired you."

"Well screw him!," said John.

"I did. You're back at work on Monday."

Advice For The Boss

1. Always leave without telling anyone where you are going. It gives me the chance to be creative when somebody asks me where you are.
2. Be nice to me only when the job I am doing for you could really change your life.
3. Do your best to keep me late. I like the office and I really don't have anything better to do.
4. If a job pleases you, keep it a secret. Leaks like that could cost me a promotion.
5. If it is really a 'rush job', run in and interrupt me every 10 minutes to inquire how it is going. That helps.
6. If my arms are full of papers, boxes, books or supplies, don't open the door for me. I need to learn how to function as a paraplegic and opening doors is good training.
7. If you don't like my work, tell everyone. I like my name to be popular in conversation.
8. If you give me more that one job to do, don't tell me which is the priority. Let me guess.
9. If you have any special instructions for a job, don't write them down. In fact, save them until the job is almost done.
10. Never give me work in the morning. Always wait until 5:00 and then bring it to me. The challenge of a deadline is refreshing.
11. Never introduce me to people you are with. When you refer to them later, my shrewd deductions will identify them.
12. Tell me all your littlest problems. No one else has any and it is nice to know that someone else is less fortunate.

Executive Level

In case you've ever wondered why ignorance rises to the executive level, here is mathematical proof:

Knowledge is Power. Time is Money. And, as every engineer knows: Power = Work / Time

If Knowledge = Power, and Time = Money, then: Knowledge = Work / Money

Solving for Money, we get: Money = Work / Knowledge

Thus, Money approaches infinity as Knowledge approaches zero, regardless of the Work done.

THEREFORE: The Less you Know, the More you Make.

Top Management Styles

BUA Management (By Using Abbreviations)
This management style is ATRASACWOC. (Adopted To Reach A Shorter And Clearer Way Of Communication)

Managing By Believing
These managers must be spiritual educated, because they have no clues at all.

Managing By Conceptual Thinking
These people try to explain the present from a theoretical view of the far future. The idea that this never will work, completely satisfies them: They will always have something to talk about.

Managing By Creating Vague Overhead Charts
Do you know them? Those charts with some big arrows, boxes or circles? These charts provide the ultimate proof of their overall brilliance.

Managing By Delegation To The Secretary
These managers just delegate everything to the secretary. If he is good, He knows what she must do.

Managing By Doing Exactly What The Boss Says
These managers prevent their bosses from creative thinking. Else they got more work to do.

Managing By Forgetting Promises
If you remind them of one of their promises, the priority of that promise is too low to remember.

Managing By Having A Non Supporting Infrastructure
In an organization with a hopeless infrastructure, managers are really necessary. These managers will naturally prevent the organization from having a better infrastructure.

Managing By Hiding Information
Information hiders are aware of the market value of strictly secret kept information. You must be very thankful to get any information at all. Beware of stimulants from category 5!

Managing By Knowing Nothing
These managers don't really know anything at all. They let YOU give answers. Meanwhile they fill the time with nice anecdotes of irrelevant cases.

Managing By Open Door And Empty Room
This is a major improvement of the older 'OPEN DOOR' management style. Now you can really walk in and out anytime you want. Nobody ever knows where these managers are.

Managing By Post-It's
> Some managers forget everything. They want to impress you with their 'busyness by continuously writing on Post-it's while you are talking.

Managing By Reorganization
> If they think there is nothing more to organize, they reorganize.

Managing By Smiling And Wearing Nice Suits
> If you drink beer with them, lunch with them, smile to them and also wear nice suits, nothing can stop your career anymore.

Managing By Speaking With Other Managers
> This kind of managing is very popular. It will give them within a few hours the same information as an employee can tell them in 15 minutes.

Managing By Staring Out Of The Window
> These managers you usually meet with their backside faced to you with their hands in their pockets. When you talk to them, their thoughts keep staring out of the windows.

Managing By Studying
> Despite their continual attendances of all kind of studies and congresses, the longer they learn, the further they get from the practice.

Managing By Using Buzz Words
> These managers like to bluff your head off with hip, nearly undefined, terms.

Managing By Walking Faster Then The Employees
> These kind of managers you will always see in the corridor, ten steps away. "We'll have to talk" you can hear them say, just as they have disappeared around the corner.

Managing By Walking One Foot Behind The Boss
> In hierarchical organizations you can watch those groups walking in the corridor. The more equal managers are directly followed by the lesser equal managers, and so on.

Signs You're Suffering From Burnout

> ➢ Visions of the upcoming weekend help you make it through Monday.
> ➢ You have so much on your mind, you've forgotten how to pee.
> ➢ You leave for a party and instinctively bring your briefcase.
> ➢ You sleep more at work than at home.

> You think about how relaxing it would be if you were in jail right now.
> You wake up to discover your bed is on fire, but go back to sleep because you just don't care.
> Your Day-Timer exploded a week ago.
> Your friends call to ask how you've been, and you immediately scream, "Get off my back, bitch!"
> Your garbage can IS your "in" box.
> You're so tired you now answer the phone, "Hell."

Excerpts From British Army Officer Evaluation Reports

> A gross ignoramus — 144 times worse than an ordinary ignoramus.
> A prime candidate for natural deselect ion.
> Got a full 6-pack, but lacks the plastic thingy to hold it all together.
> Has two brains; one is lost and the other is out looking for it.
> Bright as Alaska in December.
> Fell out of the family tree.
> If brains were taxed, he'd get a rebate.
> If you give him a penny for his thoughts, you'd get change.
> It's hard to believe that he beat out 1,000,000 other sperm.
> One neuron short of a synapse.
> Some drink from the fountain of knowledge; he only gargled.
> Was left on the Tilt-A-Whirl a bit too long as a baby.

> A photographic memory but with the lens cover glued on.
> A room temperature IQ.
> Donated his brain to science before he was done using it.
> Gates are down, the lights are flashing, but the train isn't coming.
> Got into the gene pool while the lifeguard wasn't watching.
> He's so dense, light bends around him.
> If he were any more stupid, he'd have to be watered twice a week.
> If you stand close enough to him, you can hear the ocean.
> Not the sharpest knife in the drawer.
> One-celled organisms out score him in IQ tests.
> Takes him 2 hours to watch 60 minutes.
> Wheel is turning, but the hamster is dead

Company Objective

The objective of all dedicated company employees should be to thoroughly analyze all situations; anticipate all problems prior to their occurrence; have answers for these problems; and move swiftly to solve these problems when called upon...

However, when you are up to your ass in alligators, it is difficult to remember that your initial objective was to drain the swamp.

The Boss

> A motivational sign at work: The beatings will continue until morale improves. A direct quote from the Boss: "We passed over a lot of good people to get the ones we hired."

> He's given automobile accident victims new hope for recovery. He walks, talks and performs rudimentary tasks, all without the benefit of a SPINE.

> HR Manager to job candidate "I see you've had no computer training. Although that qualifies you for upper management, it means you're under-qualified for our entry level positions."

> I thought my Boss was a bastard, and quit, to work for myself. My new Boss is a bastard, too ... but at least I respect him.

> My Boss frequently gets lost in thought. That's because it's unfamiliar territory.

> My Boss needs a surge protector. That way his mouth would be buffered from surprise spikes in his brain.

> My Boss said to me "What you see as a glass ceiling, I see as a protective barrier project!"

> Quote from a recent meeting: "We are going to continue having these meetings, everyday, until I find out why no work is getting done."

> Quote from telephone inquiry "We're only hiring one summer intern this year and we won't start interviewing candidates for that position until the Boss' daughter finishes her summer classes.

> Quote from the Boss after overriding the decision of a task force he created to find a solution: "I'm sorry if I ever gave you the impression your input would have any effect on my decision for the outcome of this"

> Quote from the Boss... "I didn't say it was your fault. I said I was going to blame it on you."

> Some people climb the ladder of success. My Boss walked under it.

Joseph J. Zajac III

Mr. Smith's Decision

Mr. Smith was looking over his books one day and decided that he wasn't making enough money to warrant two employees and he would have to lay one off. But both Sarah and Jack were such good workers he was having trouble finding a fair way to do it. He decided that he would watch them work and the first one to take a break would be the one he would lay off.

So, he sits in his office and watches them work. Suddenly, Sarah gets a terrible headache and needs to take an aspirin. She gets the aspirin out of her purse and goes to the water cooler to get something to wash it down with.

Mr. Smith follows her to the water cooler, taps her on the shoulder and says, "Sarah, I'm going to have to lay you or Jack off."

And Sarah says, "Can you jack off - I have a headache."

Job Advice

A fellow had just been hired as the new CEO of a large high tech corporation. The CEO who was stepping down met with him privately and presented him with three numbered envelopes. "Open these if you run up against a problem you don't think you can solve," he said.

Well, things went along pretty smooth, but six months later, sales took a downturn and he was really catching a lot of heat. About at his wit's end, he remembered the envelopes. He went to his drawer and took out the first envelope. The message read, "Blame your predecessor."

The new CEO called a press conference and tactfully laid the blame at the feet of the previous CEO. Satisfied with his comments, the press – and Wall Street — responded positively, sales began to pick up and the problem was soon behind him.

About a year later, the company was again experiencing a slight dip in sales, combined with serious product problems. Having learned from his previous experience, the CEO quickly opened the second envelope. The message read, "Reorganize." This he did, and the company quickly rebounded.

After several consecutive profitable quarters, the company once again fell on difficult times. The CEO went to his office, closed the door and opened the third envelope. The message said, "Prepare three envelopes."

Corporate Structure

Chairman Of The Board

- Leaps tall building in a single bound
- Is faster than a speeding bullet
- Discusses policy with God
- Is more powerful than a locomotive
- Walks on water

President

- Leaps short buildings in a single bound
- Is just as fast as a speeding bullet
- Talks with God
- Is more powerful than a switch engine
- Walks on water if the sea is calm

Executive Vice President

- Leaps short buildings with a running start and favorable winds.
- Is faster than a speeding BB.
- Is almost as powerful as a switch engine
- Talks with God if special request is approved
- Walks on water in an indoor swimming pool

Vice President

- Barely clears a quonset hut
- Can fire a speeding bullet
- Is occasionally addressed by God
- Loses tug-of-war with a locomotive
- Swims well

General Manager

- Makes high marks on the wall when trying to leap buildings
- Is run over by locomotive
- Talks to animals
- Can sometimes handle a gun without inflicting self-injury
- Dog paddles

Manager

- Runs into buildings
- Is not issued ammunition
- Recognizes locomotive two out of three times
- Talks to walls
- Cannot stay afloat with a life preserver

Trainee

- Falls over doorsteps when
- Mumbles to him/herself

trying to enter building
- ➤ Wets him/herself with a water pistol
- ➤ Says "look at the choo-choo"

- ➤ Plays in mud puddles

Secretary
- ➤ Lifts buildings and walks under them
- ➤ Catches speeding bullets in his/her teeth
- ➤ Kicks locomotives off the tracks

- ➤ Is God

- ➤ Freezes water with a single glance

Yuppies

A yuppie opened the door of his BMW, when suddenly a car came along and hit the door, ripping it off completely. When the police arrived at the scene, the yuppie was complaining bitterly about the damage to his precious BMW.

"Officer, look what they've done to my Beeeeemer!!!", he whined.

"You yuppies are so materialistic, you make me sick!!!", retorted the officer. "You're so worried about your stupid BMW, that you didn't even notice that your left arm was ripped off!!!"

"Oh my gaaawd...," replied the yuppie, finally noticing the bloody left shoulder where his arm once was, "Where's my Rolex?!!!!!"

Relationships

Becky and Jake

Becky was on her deathbed, with her husband Jake at her side. He held her cold hand and tears silently streamed down his face. Her pale lips moved. "Jake," she said.

"Hush," he quickly interrupted, "don't talk." But she insisted.

"Jake," she said in her tired voice. "I have to talk. I must confess."

"There is nothing to confess," said the weeping Jake. "It's all right. Everything's all right."

"No, no. I must die in peace. I must confess, Jake, that I have been unfaithful to you."

Jake stroked her hand. "Now Becky, don't be concerned. I know all about it," he sobbed. "Why else would I poison you?

Family Finances

A couple was having a discussion about family finances. Finally the husband exploded, "If it weren't for my money, the house wouldn't be here!"

The wife replied, "My dear, if it weren't for your money I wouldn't be here."

Translations

"It's up to you dear."
    Really means... "I am incapable of making a decision and don't really give a damn."

"You know how bad my memory is."
    Really means... "I remember the theme song to 'F Troop', the address of the first girl I ever kissed and the Vehicle Identification Numbers of every car I've ever owned, but I forgot your birthday."

"I was just thinking about you, and got you these roses."
    Really means... "The girl selling them on the corner was a real babe."

"Football is a man's game."
    Really means... "Women are generally too smart to play it."

"Oh, don't fuss. I just cut myself, it's no big deal."
    Really means... "I have actually severed a limb, but will bleed to death before I admit I'm hurt."

"I do help around the house."
    Really means... "I once put a dirty towel in the laundry basket."

"Hey, I've got my reasons for what I'm doing."
    Really means... "And I sure hope I think of some pretty soon."

"I cannot find it."
    Really means... "It didn't fall into my outstretched hands, so I'm completely clueless."

"What did I do this time?"
    Really means... "What did you catch me at?"

"What do you mean, you need new clothes?"
    Really means... "You just bought new clothes 3 years ago."

"She's one of those rabid feminists."
    Really means... "She refused to make my coffee."

Memory

A man is jogging in the park when he comes across a 98 year old man weeping on a park bench. The jogger stops to see if he is ok.

The old man replies "Life couldn't be better. I'm living with a nineteen year old nymphomaniac! In the morning when I wake up we have sex. Then she brings me breakfast in bed. After breakfast we have sex again and I have my mid-morning nap."

He continued "We normally eat out for lunch at a nice restaurant and then it's back into bed for 'afters'. Then I spend the afternoon watching sports or old movies before she cooks dinner for me..." ..."Oh, did I mention she was a gourmet chef? After dinner we have sex again and I finally collapse in bed exhausted and ready for a restful nights sleep."

Surprised, the jogger inquires. "That's my idea of bliss! Why in the world are you so upset?"

Through his tears, the old man weeps, "I cannot remember where I live!"

Heard Over Lunch

Three women are having lunch, discussing their husbands. The first says, "My husband is cheating on me, I just know it. I found a pair of stockings in his jacket pocket, and they weren't mine!"

The second says, "My husband is cheating on me, I just know it. I found a condom in his wallet, so I poked it full of holes with my sewing needle!"

The third woman fainted.

Old Couple

A little old couple in their eighties was sitting on the couch watching the Playboy movie channel. He looked at her and asked, "Do you think we can still do that?"

"Well, we can sure try!" she answered. So they shuffled off to the bedroom. He went into the bathroom to get ready and she took off all her clothes in the bedroom. When he came out of the bathroom, he saw her standing on her head in the middle of the bedroom floor.

"What are you doing, sweetheart?" he asked.

"Well," she replied, "I thought if you couldn't get it up, maybe you could just drop it in!

Fair Ride

A farmer and his wife went to a fair. The farmer was fascinated by the airplane rides but he balked at the $10 tickets.

"Let's make a deal," said the pilot. "If you and your wife can ride without making a single sound, I won't charge you anything. Otherwise you pay the ten dollars."

"Good deal!" said the farmer.

So they went for a ride. When they got back the pilot said, "If I hadn't been there, I never would have believed it. You never made a sound."

"It wasn't easy, either," said the farmer. "I almost yelled when my wife fell out."

Funeral Costs

Max dies and leaves Sadie with a total of $60,000 to her name. After everything is done at the funeral home and cemetery, she tells her closest friend that she has no money left.

The friend says, "How can that be? You told me you still had $60,000 left just a few days before Max died. How could you be broke?"

The widow says, "Well, the funeral home cost me $5,000. And of course, I had to make the obligatory donation to the temple, so that was another $5,000. The rest went for the memorial stone."

The friend says, "$40,000 for the memorial stone? How big is it?"

Extending her left hand, the bereaved widow says, "Three carats."

Observations

> ➤ A happy marriage is a matter of give and take; the husband gives and the wife takes.
> ➤ A little boy asked his father, "Daddy, how much does it cost to get married?" And the father replied, "I don't know, son, I'm still paying for it."

- A man inserted an 'ad' in the classifieds: "Wife wanted." Next day he received a hundred letters. They all said the same thing: "You can have mine."
- A perfect wife is one who helps the husband with the dishes.
- A woman was telling her friend, "It is I who made my husband a millionaire." "And what was he before you married him?" Asked the friend. The woman replied, "A multimillionaire."
- According to the latest surveys, when making love, most married men fantasize that their wives aren't fantasizing.
- Actually, should the truth be known, there are a lot of good ways to "handle" a woman. Unfortunately, not a man alive knows any of them.
- After a quarrel, a wife said to her husband, "You know, I was a fool when I married you." And the husband replied, "Yes, dear, but I was in love and didn't notice it."
- At the cocktail party, one woman said to another, "Aren't you wearing your wedding ring on the wrong finger?" The other replied, "Yes I am, I married the wrong man."
- Did any of you other married guys out there ever wonder whether it's better to have loved and lost, than to have loved and won?
- Getting married is very much like going to a restaurant with friends. You order what you want, then you see what the other fellow has, and you wish you had ordered that.
- It doesn't matter how often a married man changes his job, he still ends up with the same boss.
- Just think, if it weren't for marriage, men would go through life thinking they had no faults at all.
- Man is incomplete until he is married. Then he is really finished.
- Marriage is an institution in which a man loses his bachelor's degree and the woman gets her master's.
- Married life is very frustrating. In the first year of marriage, the man speaks and the woman listens. In the second year, the woman speaks and the man listens. In the third year, they both speak and the neighbors listen.
- Personally I think one of the greatest things about marriage is that as both husband and Father, I can say anything I want to around the house. Of course, no one pays the least bit of attention.
- Then there was a man who said, "I never knew what real happiness was until I got married; and then it was too late."
- We have a young couple in the neighborhood who are truly inseparable. Last week, it took four Howard County Policemen and a dog.
- When a man opens the door of his car for his wife, you can be sure of one thing: either the car is new or the wife.

> When a newly married man looks happy we know why. When a ten-year married man looks happy - we wonder why.
> You know the honeymoon is pretty much over when you start to go out with the boys on Wednesday nights, and so does she.
> Young son: "Is it true, Dad, I heard that in some parts of Africa a man doesn't know his wife until he marries her?" Dad: "That happens in most countries, son."

Owning a Bar

A man walks into a bar one night. He goes up to the bar and asks for a beer. "Certainly, sir, that'll be 1 cent." "ONE CENT!" exclaims the guy. The barman replies, "Yes."

So the guy glances over at the menu, and he asks, "Could I have a nice juicy T-bone steak, with chips, peas, and a fried egg?" "Certainly, sir," replies the bartender, "but all that comes to real money." "How much money?" inquires the guy. "4 cents", he replies.

"FOUR cents!" exclaims the guy. "Where's the guy who owns this place?"

The barman replies, "Upstairs with my wife." The guy says, "What's he doing with your wife?" The bartender replies, "Same as I'm doing to his business."

Calling in Sick

"Boss I not come work today I really sick. I got headache, stomach ache and my legs hurt, I not come work."

The boss says, "You know Carlos I really need you today. When I feel like this I go to my wife and tell her to give me sex. That makes me feel better and I can go to work. You should try that."

2 hours later Carlos calls back. "Boss, I did what you said and I feel great, I be at work soon. You got nice house."

Heaven

This 85 year old couple, having been married almost 60 years, had died in a car crash. They had been in good health the last ten years mainly due to her interest in health food and exercise.

When they reached the pearly gates, St. Peter took them to their mansion which was decked out with a beautiful kitchen and master bath suite and Jacuzzi. As they "oohed and aahed" the old man asked Peter how much all this was going to cost. "It's free," Peter replied, "this is Heaven."

Next they went out back to survey the championship golf course that the home backed up to. They would have golfing privileges everyday and each week, the course changed to a new one representing the great golf courses on earth.

The old man asked, "What are the green fees?" Peter's reply, "This is heaven, you play for free."

Next they went to the club house and saw the lavish buffet lunch with the cuisines of the world laid out. "How much to eat?" asked the old man. "Don't you understand yet? This is heaven, it is free!" Peter replied with some exasperation.

"Well, where are the low fat and low cholesterol tables?" the old man asked timidly. Peter lectured, "That's the best part...you can eat as much as you like of whatever you like and you never get fat and you never get sick. This is Heaven."

With that the old man went into a fit of anger, throwing down his hat and stomping on it, and shrieking wildly. Peter and the old man's wife both tried to calm him down, asking him what was wrong. The old man looked at his wife and said, "This is all your fault. If it weren't for your blasted bran muffins, I could have been here ten years ago!

Wedding Night

A young couple, just married, were in their honeymoon suite on their wedding night. As they were undressing for bed, the husband who was a big burly man tossed his pants to his bride and said, "Here, put these on."

She put them on and the waist was twice the size of her body. "I cannot wear your pants," she said. "That's right," said the husband, "and don't you ever forget it. I'm the man and I wear the pants in this family."

With that she flipped him her panties and said, "Try these on." He tried them on and found he could only get them on as far as his kneecaps. "Heck," he said, "I cannot get into your panties!"

She replied, "That's right, and that's the way it is going to be until your attitude changes!"

Three Words

A man was sitting at a bar enjoying an after-work cocktail when an exceptionally gorgeous, sexy young woman entered. She was so striking that the man could not take his eyes away from her. The young woman noticed his overly attentive stare and walked directly towards him.

Before he could offer his apologies for being so rude, the young woman said to him, "I'll do anything, absolutely anything, that you want me to do, no matter how kinky, for $100 on one condition."

Flabbergasted, the man asked what the condition was. The young woman replied, "You have to tell me what you want me to do in just three words."

The man considered her proposition for a moment, withdrew his wallet from his pocket, slowly counted out five $20 bills, which he pressed into the young woman's hand. He looked into her eyes and slowly, meaningfully said, "Paint my house."

34

Divorce

Visiting a lawyer for advice, the wife said, "I want you to help me obtain a divorce. My husband is getting a little queer to sleep with."

"What do you mean?" asked the attorney. "Does he force you to indulge in unusual sex practices?"

"No, he doesn't," replied the woman, "and neither does the little queer."

Golf Balls

A wife is cleaning out her closet and on the top shelf she notices a large box. She carefully takes the box down. She notices a sign on top of the box which reads: DO NOT OPEN!

Naturally the wife was curious so she opens the box and inside she sees $20,000 in cash and three golf balls. Later that evening her husband comes home, and she immediately confronts him about the contents of his box.

The husband is upset, but his wife proceeds, "Why are there three golf balls in the box?"

"Every time we had BAD sex I put a golf ball in the box," The husband replied.

"Hmm, three golf balls, twenty years of marriage, that's not bad," She thought. "What's the $20,000 for?" she asked.

"Every time I got a dozen golf balls, I sold them."

Sperm Bank Robbery

A guy walks into a sperm donor bank wearing a ski mask and holding a gun. He goes up to the nurse and demands her to open the sperm bank vault.

She says "But sir, its just a sperm bank!" "I don't care, open it now!!!" he replies. So she opens the door to the vault and inside are all the sperm samples.

The guy says "Take one of those sperm samples and drink it!" She looks at him "BUT they are sperm samples???" "DO IT!" So the nurse sucks it back.

"That one there, drink that one as well." So the nurse drinks that one as well. Finally after 4 samples the man takes off his ski mask. It's her husband!

He looks at her with a smile on face and says, "See honey - its not that hard."

Pet Frog

A woman went into a pet shop to buy her husband a pet. After looking around she realized that all the pets there were very expensive. She went to the counter and questioned the clerk. "I wanted to buy my husband a pet, but all of yours are so expensive."

"Well," said the clerk, "I have a huge bullfrog in the back for $50.00. Would you like to see it?" $50.00?? For a Frog??" asked the woman.

Yes, the clerk said," But it's a special frog. It gives blow jobs."

Well the woman did not particularly enjoy giving blowjobs so she thought this was a heck of a deal. She'd get her husband a gift he'd surely enjoy, and she'd never have to do that again.

The woman decided to buy the frog. She took it home to her husband and explained the strange gift. Naturally the husband was a bit skeptical, but said for sure he'd try it out that night.

The woman went to bed that night relieved, knowing she'd never have to give another blow job.

Around two in the morning, she woke up to hear pots and pans banging around in the kitchen. She got up to go see what was going on. When she arrived in the kitchen she saw her husband and the frog, sitting at the kitchen table like best buddies, looking through cookbooks.

"What are you two doing looking through cookbooks at this hour?" asked the woman.

The guy looks up at her and says, "Well, if I can teach this frog to cook, your ass is outta here.

Best Friend

A guy walked into a bar and ordered a triple scotch. The bartender poured him the drink and the guy drank it down in one gulp. "Wow," said the bartender. "Something bad musta happened."

"I came home early today," answered the guy. "I went up to the bedroom, and there was my wife having sex with my best friend." The bartender poured the dude another triple shot. "This one's on the house."

The guy gulped it down once again. The bartender asked, "Did you say anything to your wife?" The guy answered, "Yeah, I walked up to her and told her we were through. 'Pack your bag's and get out!' I told her."

"What about your friend?" asked the bartender. "I looked him straight in the eye and said, 'Bad dog!'"

Card Game

A woman was in bed having sex with her husband's best friend when all of a sudden the telephone rings and she answers. After hanging up she says, "That was my husband, but don't worry, he won't be home for awhile. He's playing cards with you."

The Date

A guy on a date parks and gets the girl in the back seat and they make love. The girl wants it again and the guy obliges her. She wants more and they do it again. She still wants more and the guy says "Excuse me a minute I have to relive myself."

While out of the car he notices a guy a half block away changing a flat. He asks the guy "Look, I've got this gal in my car and I've given it to her four or five times and she still wants more. I'll change your flat if you'll take over for me."

The guy does and is just getting in the high numbers when a cop knocks on the window and shines a light on them. The cop asks, "What're you doing in there?"

The guy says, "I'm making love to my wife."

The cop asks, "Why don't you do that at home?"

The guy answers, To tell you the truth, I didn't know it was my wife until you shined the light on her."

International Lovers

A Frenchman, an Italian and an American were discussing love-making.

"Last night I made love to my wife three times" boasted the Frenchman. "She was in sheer ecstasy this morning..."

"Ah, last night I made love to my wife six times," the Italian responded, "and this morning she made me a wonderful omelets and told me she could never love another man."

When the American remained silent, the Frenchman smugly asked, "And how many times did you make love to your wife last night?"

"Once." he replied.

"Only once?" the Italian arrogantly snorted. "And what did she say to you this morning?"

"Don't stop."

Hypothetically vs. Realistically

One day a boy comes home from school and says, "Dad, I need to know the meaning of hypothetically and realistically for school."

So the father replies, "Go ask your mother if she would sleep with a man for 1 million dollars." So the little boy goes and asks and sure enough she says yes.

His dad says ok now go ask your sister if she would sleep with a man for a million dollars. So, he does and sure enough she says yes.

So the father says, "You see son, hypothetically we are sitting on 2 million dollars but realistically, we are living with a couple of whores."

Joseph J. Zajac III

Speeding

A driver is stopped by a police officer. The driver asks, "What's the problem officer?" Officer: "You were going at least 75 in a 55 zone." Man: "No sir, I was going 65." Wife: "Oh, Harry. You were going 80." (The man gives his wife a dirty look.)

Officer: "I'm also going to give you a ticket for your broken tail light." Man: "Broken tail light? I didn't know about a broken tail light!" Wife: "Oh Harry, you've known about that tail light for weeks." (The man gives his wife another dirty look.)

Officer: "I'm also going to give you a citation for not wearing your seatbelt." Man: "Oh I just took it off when you were walking up to the car." Wife: "Oh Harry, you never wear your seatbelt." The man turns to his wife and yells, "SHUT YOUR MOUTH!"

The Officer turns to the woman and asks, "Ma'am, does your husband talk to you this way all the time?" The wife says, "No, only when he's drunk."

Canceled Trip

A man's business trip is cancelled and he is at home with a rather nervous wife. They go to bed, but about midnight, the phone rings. The man rolls over and answered... "Hello?" "What?" "How the hell should I know, I live in Phoenix."

He hangs up and his wife asks, "Who was it dear?"

"Just some idiot who wanted to know if the coast was clear!"

Yuppies

A Yuppie was accosted by a hooker. She said, "How 'bout some relaxing oral sex honey... only $50... you look all uptight."

"No way!" the man responded. "I'm married!!!"

"So???" queried the hooker.

"My wife will do it for $35." he replied.

Newlyweds

A newly married man asks his wife, "Would you have married me if my father hadn't left me a fortune?"

"Darling," the woman replies sweetly," I'd have married you no matter who left you a fortune."

Honeymooners

A new bride was a bit embarrassed to be known as a honeymooner and asked her new husband when they pulled up to the hotel if there was any way they could make it appear that they had been married a long time.

He responded: "Sure. You carry the suitcases."

He Said/She Said

He said:	I don't know why you wear a bra; you've got nothing to put in it.
She said:	You wear briefs, don't you?
She said:	What do you mean by coming home half drunk?
He said:	It's not my fault...I ran out of money.
He said:	Since I first laid eyes on you, I've wanted to make love to you in the worst way.
She said:	Well, you succeeded.
He said:	Two inches more, and I would be king'.
She said:	Two inches less, and you'd be queen'
He said:	Shall we try a different position tonight?"
She said:	"That's a good idea...you stand by the ironing board while I sit on the sofa and fart.
Priest:	I don't think you will ever find another man like your late husband.
She said:	Who's gonna look?
He said:	What have you been doing with all the grocery money I gave you?
She said:	Turn sideways and look in the mirror.
He said:	Let's go out and have some fun tonight.
She said:	Okay, but if you get home before I do, leave the hallway light on.
He said:	Why don't you tell me when you have an orgasm?
She said:	I would but you're never there.

Plane Trouble

A passenger plane on a Transatlantic flight runs into a terrible storm. The plane gets pounded by rain, hail, wind and lightning. The passengers are screaming. They are sure the plane is going to crash and they are all going to die.

At the height of the storm, a young woman jumps up and exclaims, "I cannot take this anymore! I cannot just sit here and die like an animal, strapped into a chair. If I am going to die, let me die feeling like a woman. Is there anyone here man enough to make me feel like a woman?"

She sees a hand raise in the back, and a muscular man starts to walk up to her seat. As he approaches her, he takes off his shirt. She can see the man's muscles even in the poor lighting of the plane.

He stands in front of her, shirt in hand and says to her, "I can make you feel like a woman before you die. Are you interested?"

Her eyes filled with admiration for his statuesque physique, she nods her head. "Yes!"

As the man hands her his shirt, he says, "Here, iron this."

Zoo Trip

It's a beautiful warm spring day and a man and his wife are at the zoo. She's wearing a cute, loose-fitting, pink spring dress, sleeveless w/straps. As they walk through the ape exhibit and pass in front of a very large gorilla, the gorilla goes ape. He jumps up on the bars, holding on with one hand (and 2 feet), grunting and pounding his chest w/the free hand. He is obviously excited at the pretty lady in the wavy dress.

The husband, noticing the excitement, suggests that his wife tease the poor fellow. The husband suggests she pucker her lips, wiggle her bottom, and play along. She does and Mr. Gorilla gets even more excited, making noises that would wake the dead. Then the husband suggests that she let one of her straps fall, she does, and Mr. Gorilla is just about to tear the bars down.

Now try lifting your dress up your thighs... this drives the gorilla absolutely crazy. Then, quickly the husband grabs his wife by the hair, rips open the door to the cage, slings her in with the gorilla and says, "Now, tell HIM you have a headache."

Golf Genie

A couple went golfing one day at a very exclusive course lined with million dollar homes. On the third tee, the husband cautioned, "Honey, be careful when you drive. If we break one of those windows it'll cost us a fortune to repair."

Of course, she tee'd off and promptly shanked it right through the window of the biggest house on the course. The husband cringed, "I warned you to watch out! Now we'll have to go up there and apologize and see how much that lousy drive is going to cost us."

They walked up, knocked on the door, and a warm voice said, "Come on in." When they opened the door they saw glass all over the place and a broken antique bottle lying on its side near the broken window. A man reclining on the couch asked, "Are you the people that broke the window?" "Uh yeah, we're sure sorry about that" the husband replied.

"Oh, no apology is necessary. Actually I want to thank you. You see, I'm a genie, and I've been trapped in that bottle for a thousand years. Now that you have released me, I'm allowed to grant three wishes. I'll give you each one wish, and I'll keep the last one for myself." "Wow, that's great!" the husband said.

He pondered a moment and blurted out, "I'd like a million dollars a year for the rest of my life." "No problem", said the genie, "You've got it, it's the least I can do." "And now you, young lady, what do you want?" the genie asked.

"I'd like to own a gorgeous home complete with servants in every country in the world" she said. "Consider it done." the genie said. "And now," the couple both asked in unison, "what's your wish, genie?" "Well, since I've been trapped in that bottle and haven't been with a woman in a thousand years, my wish is to have sex with your wife."

The husband looked at his wife and said, "Gee, honey, you know we both now have a fortune, and all those houses. What do you think?" She mulled it over for a few moments and said, "You know, you're right. Considering all that, I guess I wouldn't mind." The genie and the woman went upstairs where he ravished her for the rest of the afternoon.

Both satisfied each other repeatedly, and afterwards, the genie rolled over and looked at the wife and asked, "How old are you and your husband?" "Why, we are both 35" she responded breathlessly. "No shit! Thirty-five years old and both of you still believe in genies?"

Search for the Perfect Wife

Mulla Nasruden was sitting in a tea shop when a friend came excitedly to speak with him. "I'm about to get married, Mulla," his friend stated, "and I'm very excited. Mulla, have you ever thought of marriage yourself?"

Nasrudin replied, "I did think of getting married. In my youth in fact I very much wanted to do so. I waited to find for myself the perfect wife."

"I traveled looking for her, first to Damascus. There I met a beautiful woman who was gracious, kind, and deeply spiritual, but she had no worldly knowledge."

"I traveled further and went to Ishphalan. There I met a woman who was both spiritual and worldly, beautiful in many ways, but we did not communicate well."

"Finally, I went to Cairo and there after much searching I found her. She was spiritually deep, graceful, and beautiful in every respect, at home in the world and at home in the realms beyond it. I felt I had found the perfect wife."

His friend questioned further, "Then did you not marry her, Mulla?"

"Alas," said Nasrudin as he shook his head, "She was, unfortunately, waiting for the perfect husband."

Dinner Out

A married couple was enjoying a dinner out when a statuesque redhead walked over to their table, exchanged warm greetings with the husband, and walked off.

"Who was that?" the wife demanded.

"If you must know," the husband replied, "that was my mistress."

"Your mistress? That's it! I want a divorce!" the wife fumed.

The husband looked her straight in the eye and said, "Are you sure you want to give up the big house in the suburbs, the Rolls, the furs, the jewelry, and the vacation home in Mexico?"

For a long time they continued dining in silence. Finally, the woman nudged her husband and said, "Isn't that Howard over there? Who's he with?"

"That's HIS mistress," her husband replied.

"Hmmph..." she said, taking a bite of dessert. "Ours is much cuter."

Mouse that Roared

One night a man heard howls coming from his basement and went down to discover a female cat being raped by a mouse. Fascinated by what he saw, the man gained the mouse's confidence with some cheese and then took him next door.

The mouse repeated his amazing performance by raping a German Shepherd. The man, very excited by this, was dying to show someone his discovery. He rushed home and woke up his wife but before he could explain, she saw the mouse, screamed, and covered her head with the blanket.

"Don't be afraid, darling," said the man. "Wait until I tell you about this."

"Get out of here!" cried his wife. "And take that sex maniac with you!"

Rest Home Memories

It was at the Golden Age Rest Home that Sam met Gilda. He was in love! He constantly told her how beautiful she was, what a great sense of humor she had, and how sexy she was. "Oh Sammy", she said, "if you knew how old I was, you wouldn't be saying such things."

But Sam was in love. "Listen Gilda, I bet I can tell you exactly how old you are" he said. "And how would you do that?" she replied. "Well Gilda", he said, "we will go to your room, light a candle, dim the lights, get on the bed, and I will put my hand in your panties. Then I will tell you just exactly how old you are." Being the curious type, Gilda agreed.

They went into her room, lit a candle, dimmed the lights, got on the bed, and Sam put his hand inside Gilda's panties. When they were 'finished'...Gilda asked, "Ok then, how old am I?" Sam replied, "88 years old." "OY!!! I don't believe it! How did you know?" am replied, "You told me last week!"

Shorts

Wife: The 2 things I cook best are meatloaf and apple pie.
Husband: Which is this?

Wife: Do you want dinner?
Husband: Sure, what are my choices?
Wife: Yes and no.

Tee Off

A guy stood over his tee shot for what seemed an eternity; looking up, looking down, measuring the distance, figuring the wind direction and speed. Driving his partner nuts. Finally his exasperated partner says, "What's taking so long? Hit the blasted ball!"

The guy answers, "My wife is up there watching me from the clubhouse. I want to make this a perfect shot."

"Forget it, man-you don't stand a snowball's chance in hell of hitting her from here!"

The Perfect Day According to...

Her		**Him**	
8:45	Wake up to hugs & kisses	10:00	Wake up
9:00	5 pounds lighter on the scales	10:02	Oral Sex
10:00	Light breakfast	10:45	Big breakfast
11:00	Sunbathe	11:30	Drive in Ferrari with gorgeous blonde
12:00	Lunch with best friend at outdoor cafe	2:15	Enormous lunch
1:30	Shopping	3:00	Oral Sex
2:30	Run into boyfriend's ex, she's gained 30 lbs	3:30	Play sport with the guys
3:00	Facial, massage and nap	4:00	Drink beer with the guys
5:30	Talk with mom on the phone for an hour	6:00	Meet Claudia Schiffer
7:30	Candlelit dinner for two and dancing	6:10	Oral Sex
10:00	Make love	8:00	Huge dinner, more beer
11:00	Pillow talk in his big strong arms	11:00	Full on, get down, gorilla sex

> 11:30   Watch late game from
> the West Coast

Psychiatrist

A woman went to her psychiatrist because she was having severe problems with her sex life. The psychiatrist asked her many questions but did not seem to be getting a clear picture of her problems.

Finally he asked, "Do you ever watch your husband's face while you are having sex?"

"Well, yes, I did once."

"Well, how did he look?"

"Very angry."

At this point the psychiatrist felt that he was really getting somewhere and he said, "Well that's very interesting, we must look into this further. Now tell me, you say that you have only seen your husband's face once during sex; that seems somewhat unusual; how did it occur that you saw his face that time?"

"He was looking through the window."

The Dictionary Of Dating

Attraction
The act of associating horniness with a particular person.

Birth Control
Avoiding pregnancy through such tactics as swallowing special pills, inserting a diaphragm, using a condom, and dating repulsive men or spending time around children.

Dating
The process of spending enormous amounts of money, time, and energy to get better acquainted with a person whom you don't especially like in the present and will learn to like a lot less in the future.

Easy
A term used to describe a woman who has the sexual morals of a man.

Eye Contact
A method utilized by a single woman to communicate to a man that she is interested in him. Despite being advised to do so, many women have difficulty looking a man directly in the eyes, not necessarily due to the shyness, but usually due to the fact that a woman's eyes are not located in her chest.

Friend
> A member of the opposite sex in your acquaintance who has some flaw which makes sleeping with him/her totally unappealing.

Frigid
> A man's term for a woman who wants to have sex less often than he does, or who requires more foreplay than lifting her nightgown.

Indifference
> A woman's feeling towards a man, which is interpreted by the man as "playing hard to get."

Interesting
> A word a man uses to describe a woman who lets him to all the talking.

Irritating Habit
> What the endearing little qualities that initially attract two people to each other turn into after a few months together.

Law Of Relativity
> How attractive a given person appears to be is directly proportional to how unattractive your date is.

Love At 1st Sight
> What occurs when two extremely horny, but not entirely choosy people meet.

Nag
> A man's term for a woman who wants more to her life with him than just intercourse.

Nymphomaniac
> A man's term for a woman who wants to have sex more often than he does.

Prig
> A term used to describe a woman who wants to stay virgin until married.

Sober
> Condition in which it is almost impossible to fall in love.

After the Honeymoon

Typical macho man married typical good-looking lady, and after the wedding, he laid down the following rules:

"I'll be home when I want, if I want and at what time I want - and I don't expect any hassle from you. I expect a great dinner to be on the table, unless I tell you. I'll go hunting, fishing, boozing, and card playing whenever I want with my old buddies, and don't you give me a hard time about it. Those are my rules. Any comments?"

His new bride said, "No, that's fine with me. Just understand that there'll be sex here at seven o'clock every night, whether you're here or not."

40th Wedding Anniversary

Husband and wife had a bitter quarrel on the day of their 40th Wedding anniversary. The husband yells, "When you die, I'm getting you a headstone that reads, 'Here Lies My Wife - Cold As Ever.'"

"Yeah" she replies,

"When you die, I'm getting you a headstone that reads, 'Here Lies My Husband - Stiff At Last.'"

Dentist

A woman and her husband interrupted their vacation to go to the dentist. "I want a tooth pulled, and I don't want any pain killers because I'm in a big hurry," the woman said. "Just extract the tooth as quickly as possible, and we'll be on our way."

The dentist was quite impressed. "You're certainly a courageous woman," he said. "Which tooth is it?"

The woman turned to her husband and said, "Show him your tooth, dear."

Nicknames

A man has six children and is very proud of his achievement. He is so proud of himself that he starts calling his wife "Mother of Six" in spite of her objections. One night, they go to a party. The man decides that it's time to go home, and wants to find out if his wife is ready to leave as well. He shouts

At the top of his voice, "Shall we go home, Mother of six?"

His wife, irritated by her husband's lack of discretion, shouts back... "Anytime you're ready, Father of Four!"

To the Store

A man walks into a pharmacy and wanders up and down the aisles. The sales girl notices him and asks if she can help him.

He answers that he looking for a box of tampons for his wife. She directs down the correct aisle. A few minutes later, he deposits a huge bag of cotton balls on the counter.

She says, confused, "Sir, I thought you were looking for tampons for your wife?"

"You see it's like this. Yesterday, I sent my wife to the store to get me a carton of Cigarettes and she came home with a tin of tobacco and some rolling paper. So, I figure, if I have to roll my own, SO DOES SHE!"

Love, Lust, Or Really Married # 1

LOVE	When your eyes meet across a crowded room.
LUST	When your tongues meet across a crowded room.
MARRIAGE	When you lose your child in crowded room.
LOVE	When intercourse is called "making love."
LUST	When intercourse is called "screwing."
MARRIAGE	What the hell are you talking about?
LOVE	When you argue over how many children to have.
LUST	When you argue over who gets the wet spot.
MARRIAGE	When you argue over money.
LOVE	When you share everything you own.
LUST	When you steal everything they own.
MARRIAGE	When the bank owns everything.
LOVE	When it doesn't matter if you don't climax.
LUST	When the relationship is over if you don't climax.
MARRIAGE	What's a climax?
LOVE	When you phone each other just to say, "Hi."
LUST	When you phone each other to pick a hotel room.
MARRIAGE	When you phone each other to bitch.
LOVE	When you write poems about your partner.
LUST	When all you write is your phone number.
MARRIAGE	When all you write is checks.

Doctor's Wife

A doctor and his wife are having a fight at the breakfast table. The husband gets up in a rage, and says, "and you're no good in bed either," and he storms out of the house. After sometime, he realizes he was nasty, and he decides to make amends, so, he telephones her.

She comes to the phone after many rings, and the irritated husband says, "What took you so long to answer the phone"?

She says, "I was in bed."

"In bed this late, doing what?"

"Getting a second opinion."

Mother-in-Law

When met by a long procession of people led by a man with a dog, Jim asked the man,
"Who died?"
"My Mother in law."
"How?"
"The dog bit her."
"Can I borrow the dog?"
"Get in line."

20 Years

The wife found her husband sitting on the back porch crying. "What's wrong?" she asked.
"Do you remember when we were dating and your father told me that if I didn't marry you, he would send me to prison for 20 years?" he said.
"Yes" she responded, "so what?"
I would have gotten out of prison today!" he sobbed.

Top Reasons Trick-or-Treating is Better than Sex:

> ➢ 40 years from now, you'll still enjoy candy.
> ➢ Doesn't matter if kids hear you moaning and groaning.
> ➢ Guaranteed to get at least a little something in the sack.
> ➢ If you don't get what you want, you can always go next door!!!
> ➢ If you get tired, wait 10 minutes and go at it again.
> ➢ If you wear a Bill Clinton mask, no one thinks you're kinky.
> ➢ Less guilt the next morning.
> ➢ The person you're with doesn't fantasize you're someone else.
> ➢ The uglier you look, the easier it is to get some.
> ➢ You don't have to compliment the person who gave you candy.

Fur Coat

A man and a woman walk into a very posh Rodeo Drive furrier. "Show the lady your finest mink!" the fellow exclaims. So the owner of the shop goes in back and comes out with an absolutely gorgeous full-length coat.
As the lady tries it on, the furrier sidles up to the guy and discreetly whispers, "Ah, sir, that particular fur goes for $65,000."
"No problem! I'll write you a check!"
"Very good, sir." says the shop owner. "Today is Saturday. You may come by on Monday to pick it up, after the check has cleared."

So the man and the woman leave. On Monday, the fellow returns. The store owner is outraged: "How dare you show your face in here?! There wasn't a single penny in your checking account!!"

"I just had to come by," grinned the guy, "to thank you for the most wonderful weekend of my life!"

Love, Lust, Or Really Married # 2

LOVE	When you show concern for your partner's feelings.
LUST	When you couldn't give a shit.
MARRIAGE	When your only concern is what's on TV.

LOVE	When your farewell is "I love you, darling..."
LUST	When your farewell is "So, same time next week..."
MARRIAGE	When your farewell is a relief.

LOVE	When you are proud to be seen in public with your partner.
LUST	When you only see each other naked.
MARRIAGE	When you never see each other awake.

LOVE	When your heart flutters every time you see them.
LUST	When your groin twitches every time you see them.
MARRIAGE	When your wallet empties every time you see them.

LOVE	When nobody else matters.
LUST	When nobody else knows.
MARRIAGE	When everybody else matters and you don't care who knows.

LOVE	When all the songs on the radio describe exactly how you feel.
LUST	When the song on the radio determines how you do it.
MARRIAGE	When you listen to talk radio.

LOVE	When breaking up is something you try not to think about.
LUST	When staying together is something you try not to think about.
MARRIAGE	When just getting through today is your only thought.

LOVE	When you're only interested in doing things with your partner.
LUST	When you're only interested in doing things TO your partner.
MARRIAGE	When you're only interested in your golf score.

Video Tape

A woman finds out that her husband is cheating on her while stationed in Saudi Arabia. So she sends him a very special care package. He is very excited to get a package from his wife back home. He finds that it contains a batch of home made cookies and a VHS tape of his favorite TV shows.

He invites a couple of his buddies over and they're all sitting around having a great time eating the cookies and watching some episodes of South Park. Right in the middle of one episode the tape cuts to a home video of his wife on her knees giving his best friend oral sex.

After a few seconds, he does his business in her mouth and she turns and spits the load right into the mixing bowl of cookie dough. She then looks at the camera and says, "By the way, I want a divorce."

Marriage Sayings

1. A man rushes into his house and yells to his wife, "Martha, pack up your things! I just won the California lottery!" Martha replies, "Shall I pack for warm weather or cold?" The man responds, "I don't care. Just as long as you're out of the house by noon!"
2. A man was complaining to a friend: "I had it all - money, a beautiful house, a big car, the love of a beautiful woman; Then, Pow! it was all gone!" "What happened?" asked the friend. "My wife found out..."
3. A man without a woman is like a fish without a bicycle.
4. How many men does it take to open a beer? None. It should be opened by the time she brings it to the couch.
5. I haven't spoken to my wife for 18 months-I don't like to interrupt her.
6. I married Miss Right. I just didn't know her first name was Always.
7. If your wife and a lawyer were drowning and you had to choose which to save, would you go to lunch or to a movie?
8. It's not true that married men live longer than single men. It only seems longer.
9. Losing a wife can be hard. In my case, it was almost impossible.
10. Wife: Let's go out and have some fun tonight. Husband: Okay, but if you get home before I do, leave the hallway light on.
11. Women will never be equal to men until they can walk down the street bald and still think they are beautiful!
12. A man is incomplete until he is married. After that, he is finished.

Make Her Scream

What's the best way to make your wife scream when you're having sex?

Call her up and tell her where you are.

Husband Passing Away

Mary Clancy goes up to Father O'Grady after his Sunday morning service, and she's in tears.

He says, "So what's bothering you, dear?"

She says, "Oh, Father, I've got terrible news. My husband passed away last night."

The priest says, "Oh, Mary, that's terrible. Tell me, Mary, did he have any last requests?"

She says, "That he did, Father..."

The priest says, "What did he ask, Mary?"

She says, "He said, 'Please, Mary, put down that gun!!!'"

On the Farm

A little boy comes down to breakfast. Since they live on a farm, his mother asks if he had done his chores. "Not yet," said the little boy. His mother tells him no breakfast until he does his chores.

Well, he's a little pissed, so he goes to feed the chickens, and he kicks a chicken. He goes to feed the cows, and he kicks a cow. He goes to feed the pigs, and he kicks a pig. He goes back in for breakfast and his mother gives him a bowl of dry cereal.

"How come I don't get any eggs and bacon? Why don't I have any milk in my cereal?" he asks. "Well," his mother says, "I saw you kick a chicken, so you don't get any eggs for a week. I saw you kick the pig, so you don't get any bacon for a week either. I also saw you kick the cow, so for a week you aren't getting any milk."

Just then, his father comes down for breakfast and kicks the cat half way across the kitchen. The little boy looks up at his mother with a smile, and says, "Are you going to tell him, or should I?"

Religion

Adam and Eve

After a few days, the Lord called Adam to him, and said, "It is time for you and Eve to begin the process of populating the Earth, so I want you to start by kissing Eve."

Adam answered, "Yes Lord, but what's a 'kiss'?" So the Lord gave Adam a brief description and Adam then took Eve by the hand, behind a nearby bush. A few minutes later, Adam emerged, and said, "Lord, that was enjoyable."

And the Lord replied, "Yes, Adam, I thought you'd enjoy that, and now I'd like you to caress Eve."

And Adam said, "Lord, what's a 'caress'?" So the Lord gave Adam a brief description and Adam went again behind the bush with Eve. Quite a few minutes later, Adam returned, smiling, and said, "Lord, that was even better than the kiss."

And the Lord said, "You've done well, Adam, and now I want you to make love to Eve."

And Adam said, "Lord, what's 'making love'?" So the Lord again gave Adam directions, and Adam went to Eve, behind the bush. But this time he reappeared in two seconds. And Adam said, "Lord, what's a 'headache'?"

Government and God

God Created heaven and the earth. Quickly he was faced with a class action suit for failure to file an environmental impact statement. He was granted a temporary permit for the project, but was stymied with the cease and desist order for the earthly part.

Appearing at the hearing, God was asked why he began his earthly project in the first place. He replied that he just liked to be creative.

Then God said, "Let there be light", and immediately the officials demanded to know how the light would be made. Would there be strip mining? What about thermal pollution? God explained that the light would come from a huge ball of fire. God was granted provisional permission to make light, assuming that no smoke would result from the ball of fire: that he would obtain a building permit; and to conserve energy, would have the light out half the time. God agreed and said he would call the light "Day" and the darkness "Night." Officials replied that they were not interested in semantics.

God said, "Let the earth bring forth green herb and such as many seed." The EPA agreed so long as native seed was used. Then God said, "Let waters bring forth creeping creatures having life; and the fowl that may fly over the earth." Officials pointed out this would require approval from the Department of Game coordinated with the Heavenly Wildlife Federation and the Audubongelic Society.

Everything was O.K. until God said he wanted to complete the project in Six days. Officials said it would take at least 200 days to review the application and impact statement. After that there would be a public hearing. Then there would be 10-12 months before...

At this point God created Hell.

Six Days

Once upon a time in the Kingdom of Heaven, God was missing for six days. Eventually, Michael the archangel found him, resting on the seventh day. He inquired of God, "Where have you been?"

God sighed a deep sigh of satisfaction and proudly pointed downwards through the clouds, "Look Michael, look what I've made." Archangel Michael looked puzzled and said, "What is it?" "It's a planet," replied God, "and I've put LIFE on it. I'm going to call it Earth and it's going to be a great place of balance."

"Balance?" inquired Michael, still confused.

God explained, pointing to different parts of Earth, "For example, Northern Europe will be a place of great opportunity and wealth while Southern Europe is going to be poor; the Middle East over there will be a hot spot. Over there I've placed a continent of white people and over there is a continent of black people," God continued, pointing to different countries. "This one will be extremely hot and arid while this one will be very cold and covered in ice."

The Archangel, impressed by Gods work, then pointed to a large land mass and said "What's that one?"

"Ah," said God. "That's New England, the most glorious place on Earth. There are beautiful lakes, rivers, streams, and mountains. The people from New England are going to be modest, intelligent and humorous and they're going to be found traveling the world. They'll be extremely sociable, hardworking and high-achieving, and they will be known throughout the world as diplomats and carriers of peace."

Michael gasped in wonder and admiration but then proclaimed, "What about balance, God? You said there will be BALANCE!"

God replied wisely, "Wait until you see the loudmouthed bums I'm putting next to them in New York."

Decisions

When God was creating the human race, he lined up all the males on one side and all the females opposite. Then he asked, "Which of your species would like to urinate standing up?"

Well, the males went crazy, shouting that they wanted to pee standing up.

"Fine", says God, "Women get multiple orgasms"

Reasons why beer is better than religion:

1. Beer doesn't tell you how to have sex.
2. Beer has never caused a major war.
3. If you've devoted your life to Beer, there are groups to help you stop.
4. No one will kill you for not drinking Beer.
5. Nobody's ever been burned at the stake, hanged, or tortured over his brand of Beer.
6. There are laws saying Beer labels cannot lie to you.
7. They don't force Beer on minors who cannot think for themselves.
8. When you have a Beer, you don't knock on people's doors trying to give it away.
9. You can prove you have a Beer.
10. You don't have to wait 2000+ years for a second Beer.

Problem Parrots

This lady approaches a priest and tells him, "Father, I have a problem. I have these two talking female parrots, but they only know how to say one thing."

"What do they say?", the priest asked.

"They only know how to say `Hi, we are prostitutes. Do you want to have some FUN?'"

"That's terrible!", the priest exclaimed, "But I have a solution to your problem. Bring your two talking female parrots over to my house and I will put them with my two male talking parrots who I have taught to pray and read the Bible, then my parrots will teach your parrots to stop saying that terrible phrase and your female parrots will learn to pray and worship."

"Thank you," said the lady.

So the next day, the lady brings her female parrots to the priest's house. The priest's two male parrots are holding rosary beads and praying in their cage.

The lady puts her female talking parrots in with the male talking parrots and the female parrots say, "Hi, we are prostitutes! Do you want to have some FUN?"

One male parrot looks over to the other male parrot and says, "PUT THE BIBLES AWAY! OUR PRAYERS HAVE BEEN ANSWERED!!!!!"

Sisters of Mercy

A man is driving down a deserted stretch of highway, when he notices a sign out of the corner of his eye. It reads SISTERS OF MERCY HOUSE OF PROSTITUTION - 10 MILES. He thinks it was just a figment of his imagination and drives on without a second thought.

Soon, he sees another sign that says SISTERS OF MERCY HOUSE OF PROSTITUTION - 5 MILES and realizes that these signs are for real. When he drives past a third sign saying SISTERS OF MERCY HOUSE OF PROSTITUTION NEXT RIGHT, his curiosity gets the best of him and he pulls into the drive.

On the far side of the parking lot is a somber stone building with a small sign next to the door reading SISTERS OF MERCY. He climbs the steps and rings the bell.

The door is answered by a nun in a long black habit who asks, "What may we do for you, my son?" He answers, "I saw your signs along the highway, and was interested in possibly doing business."

"Very well, my son. Please follow me." He is led through many winding passages and is soon quite disoriented. The nun stops at a closed door, and tells the man, "Please knock on this door."

He does as he is told and this door is answered by another nun in a long habit and holding a tin cup. This nun instructs, "Please place $50 in the cup, then go through the large wooden door at the end of this hallway."

He gets $50 out of his wallet and places it in the second nun's cup. He trots eagerly down the hall and slips through the door, pulling it shut behind him. As the door locks behind him, he finds himself back in the parking lot, facing another small sign: GO IN PEACE, YOU HAVE JUST BEEN SCREWED BY THE SISTERS OF MERCY.

Three Pastors

Three Pastors in the south were having lunch in a diner.

One said "Ya know, since summer started I've been having trouble with bats in my loft and attic at church. I've tried everything—noise, spray, cats—nothing seems to scare them away.

Another said "Yea, me too. I've got hundreds living in my belfry and in the narthex attic. I've even had the place fumigated and they won't go away.

The third said, "I baptized all mine and made them members of the church... Haven't seen one back since" !!!

PMS

A preacher was telling his congregation that anything they could think of, old or new, was discussed somewhere in the Bible and that the entirety of the human experience could be found there.

After the service, he was approached by a woman who said, "Preacher, I don't believe the Bible mentions PMS." The preacher replied that he was sure it must be in there somewhere and that he would look for it.

The following week after service, the preacher called the woman aside and showed her a passage which read,

"And Mary rode Joseph's ass all the way to Bethlehem."

What Jesus Drives

Most people assume WWJD is for "What would Jesus do?" But the initials really stand for "What would Jesus drive?"

One theory is that Jesus would tool around in an old Plymouth because the Bible says, "God drove Adam and Eve out of the Garden of Eden in a Fury."

But in Psalm 83, the Almighty clearly owns a Pontiac and a Geo. The passage urges the Lord to "pursue your enemies with your Tempest and terrify them with your Storm."

Perhaps God favors Dodge pickup trucks, because Moses' followers are warned not to go up a mountain "until the Ram's horn sounds a long blast."

Some scholars insist that Jesus drove a Honda but didn't like to talk about it. As proof, they cite a verse in St. John's gospel where Christ tells the crowd, "For I did not speak of my own Accord..."

Meanwhile, Moses rode an old British motorcycle, as evidenced by a Bible passage declaring that "the roar of Moses' Triumph is heard in the hills."

Joshua drove a Triumph sports car with a hole in its muffler: "Joshua's Triumph was heard throughout the land."

And, following the Master's lead, the Apostles car pooled in a Honda.... "The Apostles were in one Accord."!!!!

Adam and Eve

A Briton, a Frenchman and a Russian are viewing a painting of Adam and Eve frolicking in the Garden of Eden. "Look at their reserve, their calm," muses the Brit. "They must be British."

"Nonsense," the Frenchman disagrees. "They're naked, and so beautiful. Clearly, they are French."

"No clothes, no shelter," the Russian points out, "they have only an apple to eat, and they're being told this is paradise. They are Russian."

Car Accident

A rabbi and a priest get into a car accident and it's a bad one. Both cars are totally demolished, but, amazingly, neither of the clerics is hurt. After they crawl out of their cars, the rabbi sees the priest's collar and says, "So you're a priest. I'm a rabbi. Just look at our cars. There's nothing left, but we are unhurt. This must be a sign from God. God must have meant that we should meet and be friends and live together in peace the rest of our days."

The priest replies, "I agree with you completely. This must be a sign from God."

The rabbi continues, "And look at this. Here's another miracle. My car is completely demolished but this bottle of Mogen David wine didn't break.

Surely God wants us to drink this wine and celebrate our good fortune." Then he hands the bottle to the priest.

The priest agrees, takes a few big swigs, and hands the bottle back to the rabbi. The rabbi takes the bottle, immediately puts the cap on, and hands it back to the priest. The priest asks, "Aren't you having any?"

The rabbi replies, "No...I think I'll wait for the police."

Confessions

A priest and a rabbi were talking when the rabbi asked the priest what goes on in the confessional. "I have an idea," said the priest. Why don't you sit with me on my side of the confession booth and hear it for yourself? No one will ever know."

A woman came into the booth and said, "Bless me Father for I have sinned." The priest asked, "What did you do?" "I cheated on my husband" she said. "How many times?" asked the priest.

"Three times," she replied. "Well," said the priest, "Say 5 Hail Marys and put 5 dollars in the offering box."

Another woman came and said, "Bless me Father for I have sinned." The priest asked, "What did you do?" "I cheated on my husband," she replied. "How many times?" asked the priest.

"Three times," she replied. Again the priest said, "Say 5 Hail Marys and put 5 dollars in the offering box."

Then the priest said to the rabbi, "would you like to do the next confession?" The rabbi started to object, but the priest said, "Go ahead. It's easy."

So another woman came in and said, "Bless me Father for I have sinned." This time the rabbi asked, "What did you do?" "I cheated on my husband," replied the woman. "How many times?" the rabbi inquired.

The woman said, "Twice." Then the rabbi said, "Well go do it again. They're 3 for 5 dollars today."

Two Nuns

Two nuns have just arrived to USA by boat and one says to the other, "I hear that the occupants of this country actually eat dogs."

"Odd," her companion replies, "but if we shall live in America, we might as well do as the Americans do." Nodding emphatically, the mother superior points to a hot dog vendor and they both walk towards it. "Two dogs, please," says one.

The vendor is only too pleased to oblige and he wraps both hot dogs in foil. Excited, the nuns hurry over to a bench and begin to unwrap their 'dogs.' The mother superior is first to open hers, then, staring at it for a moment, leans over to the other nun and whispers cautiously, "What part did you get?"

Three Nuns

Three nuns were talking. The first nun said, "I was cleaning in the father's room the other day and guess what I found? A bunch of pornographic magazines."

"What did you do?" the other nuns asked. "Of course I threw them in the trash." The second nun said, "I can top that. I was in the father's room putting away the laundry and I found a bunch of condoms!"

"Oh my!" gasped the other nuns. "What did you do?" they asked. "I poked holes in all of them!" she replied. The third nun fainted.

Three Nuns Go To Heaven

Three Italian nuns die and go to heaven, where they are met at the Pearly Gates by St. Peter. He says, "Ladies, you all led such wonderful lives, that I'm granting you six months to go back to Earth and be anyone you want."

The first nun says, "I want-a to be Sophia Loren" and <poof!> she's gone. The second says, "I want-a to be Madonna" and <poof!> she's gone. The third says, "I want-a to be Sara Pipalini."

St. Peter looks perplexed. "Who?" he says. "Sara Pipalini" replies the nun. St. Peter shakes his head and says "I'm sorry but that name just doesn't ring a bell."

The nun then takes a newspaper out of her habit and hands it to St. Peter. He reads the paper and starts laughing. He hands it back to her and says "No Sister, this says 'Sahara Pipeline laid by 500 men in 7 days'!"

Carload of Nuns

Cop pulls over a carload of nuns.
Cop: "Sister, this is a 65 MPH highway — why are you going so slow?"
Sister "Sir, I saw a lot of signs that said 22, not 65."
Cop "Oh sister, that's not the speed limit, that's the name of the highway you're on!
Sister: Oh! Silly me! Thanks for letting me know. I'll be more careful.
At this point the cop looks in the back seat where the other nuns are shaking and trembling.
Cop: Excuse me, Sister, what's wrong with your friends back there? They're shaking something terrible.
Sister: Oh, we just got off of highway 119

Lottery

A man called Jacob finds himself in dire trouble. His business has gone bust and he's in serious financial trouble. He's so desperate that he decides to ask God for help. He goes into the synagogue and begins to pray: "God, please help me. I've lost my business and if I don't get some money, I'm going to lose my house as well. Please let me win the lotto."

Lotto night comes and somebody else wins.

Jacob goes back to the synagogue: "God please let me win the lotto. I've lost my business, my house and I'm going to lose my car as well."

Lotto night comes and Jacob still has no luck!

Back to the synagogue: "My God, why have you forsaken me? I've lost my business and my house and my car and now my wife and children are starving. I don't often ask for your help and I have always been a good servant to you. Why won't you just let me win the lotto this one time so I can get my life back in order?"

Suddenly there is a blinding flash of light as the heavens open and Jacob is confronted by the booming voice of God himself who admonishes him: "Damn it Jacob, meet me half way on this one, at least buy a ticket!"

Train Ride

A priest and a Rabbi found themselves sharing a compartment on a train. After a while, the priest opened a conversation by saying "I know that, in your religion, you're not supposed to eat pork...Have you actually ever tasted it?

The Rabbi said, "I must tell the truth. Yes, I have, on the odd occasion."

Then the Rabbi had his turn of interrogation. He asked, "Your religion, too...I know you're supposed to be celibate. But..."

The priest replied, "Yes, I know what you're going to ask. I have succumbed once or twice."

There was silence for a while. Then the Rabbi peeped around the newspaper he was reading and said, "Better than pork, isn't it?"

Rest

In the beginning, God created earth and rested. Then God created man and rested. Then God created woman. Since then, neither God nor man has rested.

"Psalm 129"

A priest was driving along and saw a nun on the side of the road he stopped and offered her a lift which she accepted. She got in and crossed her legs, forcing the habit to open and reveal a leg. The priest looks and nearly has an accident and after changing gear lets his hand slide up her leg. She immediately says "Father remember Psalm 129."

The priest apologizes profusely and removes his hand but is unable to remove his eyes from her leg. Further on when he changes gear and has ogled at her leg for the zillionth time he lets the hand slide up the leg again. The Nun once again says "Father remember Psalm 129."

Once again the priest apologizes "Sorry sister but you know the flesh is weak."

Arriving at the convent the nun gets out and the priest goes on his way.

Once he arrives at his church he rushes to the bible and looks up psalm 129 it said: "GO FORTH AND SEEK, FURTHER UP YOU WILL FIND GLORY"

Moral of the story: you should always be well informed in your job or you might miss a great opportunity.

Sex Lecture

A minister gave a talk to the Lion's Club on sex. When he got home he couldn't tell his wife that he had spoken on sex, so he said he had discussed horseback riding with the members.

A few days later, she ran into some men at the shopping center and they complimented her on the speech her husband had made.

She said, "Yes, I heard. I was surprised about the subject matter, as he's only tried it twice. The first time he got so sore he could hardly walk, and the second time he fell off."

Guess Who

A local preacher was paying a visit to one of his church members on a Friday night, and heard a loud party as he approached the house.

He knocked on the door and the owner answered. Behind him, he saw a circle of naked men, with blindfolded women moving from man to man, fondling each man's package, and guessing who it was.

The preacher, seeing this, said, "I'm sorry. I don't think I'd fit in here right now."

"Nonsense," the man replied. "Your name's been called three times."

Three Monks

Three monks are deep in the Himalayas, deeply meditating. One year passes in silence, then one of them says, "Pretty cold here." Another year passes and the second monk says, "You're right. It is cold." Another year goes by and the third monk says, "If you two don't stop your bitching, I'm gonna leave."

A Comparative Guide To Religions

➤	Agnosticism	Shit may or may not happen.
➤	Apathism	I don't care if shit happens.
➤	Atheism	There is no shit.
➤	Baha'i	Shit happens to everyone equally.
➤	Baptist	Gonna wash this shit right outta my

hair!

- Branch Davidianism — This shit really burns me up.
- Buddhism — If shit happens, it really isn't shit.
- Calvinism — Shit happens because you don't work hard enough.
- Cartesianism — I think, therefore shit happens.
- Catholicism — Shit happens because you are BAD.
- Christian Science — Shit only happens within your mind.
- Church of England — Excrement transpires.
- Confucianism — Confucius say, "shit happens."
- Episcopalianism — Shit happened to my ancestors.
- Evangelism — Will you accept shit happening into your heart?
- Existentialism — What is shit, anyway?
- Hedonism — There is nothing like a good shit happening.
- Hinduism — This shit happened before.
- Islam — If shit happens, it is the will of Allah.
- Judaism — Why does shit always happen to me?
- Lutheranism — Shit happens; God help me, I can do no other.
- Methodism — Shit happens; support group available.
- Nihilism — No shit!
- Presbyterianism — This shit happens to be too dear.
- Protestantism — Let shit happen to somebody else.
- Puritanism — Shit happens with our permission.
- Quaker — Get thy shit together, Friend.
- Rastafarianism — Let's smoke this shit.
- Satanism — Shit is the best part.
- Scientific Methodism — Skatole-bearing anomalies materialize.
- Scientology — Does shit happen? (See page 469).
- Shintoism — Farts happen.
- Seventh Day Adventism — No shit happens on Saturdays.
- Stoicism — This shit happens to be good for me.
- Taoism — Shit happens.
- Televangelism — Send all your money or shit will happen.
- Unitarianism — Shit happens; discussion at 2:00 pm.
- Zen — What is the sound of one shit happening?

Politics

"A billion here, a billion there, sooner or later it adds up to real money."
- Everett Dirksen

Rejected Democratic National Committee Fundraising Ideas 1992 - 2000

1. "Hi, this is your President, for Weight Watchers..."
2. $10,000 a plate "Pant less Dinner."
3. $50,000 gets you dinner at the White House with international dignitaries. $50 gets you breakfast at the International House of Pancakes with Roger.
4. 1-900-PREZ-SEX
5. 1997 Women of the Justice Department swimsuit calendar.
6. A good pin-the-$100-bill-on-Jesse-Helms game should do the trick.
7. Al's hair declared endangered species to solicit matching federal contributions.
8. Charter Air Force One for Mile High Club initiations.
9. For $5000, Madeleine Albright will talk dirty to you in 12 different languages.
10. Get "protection money" from Nike to guarantee that the President will *not* be seen jogging in their clothing.
11. Hillary's "Better-Then-Neiman-Marcus" cookie recipe.
12. Janet Reno's "Total Amazon Workout" videos.
13. Senator Ted takes on Yeltsin in a vodka shot duel — if Boris loses, he ponies up a cool billion!
14. Time to get that Warren Christopher kissing booth a-crankin'!
15. Year 'round in the Oval Office — sit on Santa's lap for $2500.

Politically Correct Insults

- A few beers short of a six-pack
- A few feathers short of a whole duck
- A few peas short of a casserole
- As smart as bait
- His skylight leaks a little.
- The cheese slid off her cracker.
- A few clowns short of a circus
- A few fries short of a Happy Meal
- All foam, no beer
- There's no grain in the silo.
- Doesn't have all her dogs on one leash
- Her chimney's clogged.

> Her sewing machine's out of thread.
> If she had another brain, it would be lonely.
> One taco short of a combination plate
> Several nuts short of a full pouch
> Her antenna doesn't pick up all the channels.
> His elevator doesn't go all the way to the top floor.
> Couldn't pour water out of a boot with instructions on the heel

> His belt doesn't go through all the loops.
> Missing a few buttons on his remote control
> Proof that evolution can go in reverse
> Surfing in Nebraska
> The wheel's spinning, but the hamster's dead.
> There's too much yardage between the goal posts.
> Doesn't have all her Corn Flakes in one box.

Reckless Driving

Emo Phillips was pulled over in Massachusetts for reckless driving. When brought before the judge, Emo was asked if he knew what the punishment for drunk driving in that state was.

His reply: "I don't know, reelection to the Senate?"

Warranty Card On Purchased Government Official

Dear Special Interest,

Congratulations on the purchase of your genuine Government Official. With regular maintenance your Government Official[tm] should provide you with a lifetime of sweetheart deals, insider information, preferential legislation and other fine services.

Before you begin using your product, we would appreciate it if you would take the time to fill out this customer service card. This information will not be sold to any other party, and will be used solely to aid us in better fulfilling your future needs in political influence.

1. Which of our fine products did you buy? (Please check all that apply.)
__ President
__ Vice-President
__ Senator
__ Congressman
__ Governor
__ Cabinet Secretary - Commerce
__ Cabinet Secretary - Other
__ Other Elected Official (please specify)
__ Other Appointed Official (please specify)

2. How did you hear about your Government Official? (Please check all that apply.)

___ Arkansas crony of.
___ Attempted to seduce me.
___ Former law partner of.
___ Frequently mentioned in conspiracy theories. (Elsewhere.)
___ Frequently mentioned in conspiracy theories. (On Internet.)
___ Magazine / newspaper ad.
___ Procured for.
___ Recommended by lobbyist.
___ Recommended by organized crime figure.
___ Related to.
___ Shared jail cell with.
___ Solicited bribe from me.
___ Spoke at fundraiser at my temple.
___ TV ad.
___ Unindicted co-conspirator with.

3. How do you expect to use your Government Official? (Please check all that apply.)

___ Forestall military action against self / allies.
___ Have embargo imposed on enemy / rival nation / religious infidels.
___ Have embargo lifted from own nation / ally.
___ Have my prejudices turned into law.
___ Impede criminal / civil investigation of self / associates / spouse.
___ Inflict punitive legislation on class enemies / rivals / hated ethnic groups.
___ Inflict punitive regulation on business competitors / environmental exploiters / capitalist pigs.
___ Instigate military action against internal enemies / aggressors / targets for future conquest.
___ Obtain diplomatic concessions.
___ Obtain lucrative government contracts.
___ Obtain pardon for self / associates / spouse.
___ Obtain patronage job for self / spouse / mistress.
___ Obtain trade concessions.

4. What factors influenced your purchase? (Please check all that apply.)

___ Actual beliefs of Government Official.
___ Appearance.
___ Blackmail.
___ Celebrity endorsement.
___ Orders from boss / superior officer / foreign government.
___ Party affiliation.
___ Performance of currently owned model.
___ Price.

__ Professed beliefs of Government Official.
__ Reputation.

5. Is this product intended as a replacement for a currently owned Government Official? Yes/No
If you answered "yes," please indicate your reason(s) for changing models. (Please check all that apply.)
__ Convicted.
__ Dead.
__ Defect in current model:
__ Excessive operating / maintenance costs.
__ Indicted.
__ Needs have grown beyond capacity of current model.
__ Out bribed by competing interest.
__ Resigned in disgrace.
__ Senile.
__ Switched parties / beliefs.

Thank you for your valuable time. Always remember: in choosing a Government Official you have chosen the best politician that money can buy.

Politically Correct Women

Old	New
➢ Her breasts will never sag.	➢ They will lose their vertical hold.
➢ She does not cut you off.	➢ She becomes horizontally inaccessible.
➢ She does not get drunk.	➢ She becomes verbally dyslexic.
➢ She does not get drunk.	➢ She is accidentally over-served.
➢ She does not get PMS.	➢ She becomes hormonally homicidal.
➢ She does not hate sports on TV.	➢ She is athletically biased.
➢ She does not have a great butt.	➢ She is gluteus to the maximums.
➢ She does not have a hard body.	➢ She is anatomically inflexible.
➢ She does not have a killer body.	➢ She is terminally attractive.
➢ She does not have big hair.	➢ She is overly aerosoled.
➢ She does not have big	➢ Her cups runneth over.

hooters.

➢ She does not have a great rack.
➢ She does not have sexy lips.
➢ She does not shave her legs.
➢ She does not shop too much.
➢ She does not snore.
➢ She does not sunbathe.

➢ She does not wear too much make-up.
➢ She does not work out too much.
➢ She is not a bad cook.
➢ She is not a bad driver.

➢ She is not a gossip.
➢ She is not a perfect 10.
➢ She is not a screamer or moaner.
➢ She is not cold or frigid.
➢ She is not easy.
➢ She is not hooked on soap operas.
➢ She is not too skinny.
➢ She will never gain weight.
➢ You do not ask her to dance.

➢ Her breasts are centrally located.
➢ She is collagen dependent.

➢ She experiences temporary stubble reduction.
➢ She is just susceptible to marketing ploys.
➢ She is nasally repetitive.
➢ She experiences solar enhancement.
➢ She is cosmetically over saturated.
➢ She is an abdominal overachiever.
➢ She is microwave compatible.
➢ She is automotively challenged.

➢ She is a verbal terminator.
➢ She is numerically superior.
➢ She is vocally appreciative.

➢ She is thermally incompatible.
➢ She is horizontally accessible.
➢ She is melodramatically fixated.
➢ She is skeletally prominent.
➢ She will become a metabolic underachiever.
➢ You request a pre-coital rhythmic experience.

Train Ride

In a train carriage there was Bill Clinton, George Bush, a spectacular looking blonde and a frightfully awful looking fat lady. After several minutes of the trip, the train passes through a dark tunnel, and the unmistakable sound of a slap is heard. When they leave the tunnel, Clinton had a big red slap mark on his cheek.

The blonde thought: "That rascal Clinton wanted to touch me and by mistake, he must have put his hand on the fat lady, who in turn must have slapped his face"

The fat lady thought: "That dirty old Bill Clinton laid his hands on the blonde and she smacked him."

Bill Clinton thought: "George put his hand on that blonde and by mistake she slapped me."

George Bush thought: "I hope there's another tunnel soon so I can smack Clinton again."

White House Ghosts

George W. Bush was thrilled at finally being able to sleep at the White House, but something very strange happened. On the first night he was awakened by George Washington's ghost. Welcoming the opportunity to communicate with the father of our country, Bush asked, "President Washington, what is the best thing that I can do for our country?" "Set an honest and honorable example just as I did," he replied.

Later that night the ghost of Thomas Jefferson appeared and Bush asked him the same question. "Cut taxes and reduce the size of the government," Jefferson replied.

Still later the ghost of Abraham Lincoln appeared and Bush asked once again, "Abe, please tell me, what is the best thing I can do to help the nation?" Lincoln replied, "Go to the theater and see a play."

Politically Correct Terms

> ➢ Alive = Temporarily metabolically able
> ➢ Bald = Follicularly challenged
> ➢ Body Odor = Non discretionary fragrance
> ➢ Dead = Living impaired
> ➢ Dishonest = Ethically disoriented
> ➢ Fat = Horizontally challenged
> ➢ Ignorant = Knowledge-based non possessor
> ➢ Pregnant = Parasitically oppressed
> ➢ Serial-Killer = Person with difficult-to-meet needs
> ➢ Unattractive = Cosmetically different

Bus Crash

A bus load of politicians were driving down a country road when all of a sudden, the bus ran off the road and crashed into a tree in an old farmer's field. The old farmer, after seeing what happened, went over to investigate. He then proceeded to dig a hole and bury the politicians.

A few days later, the local sheriff came out, saw the crashed bus, and asked the farmer where all the politicians had gone. The old farmer said he had buried them. The sheriff then asked the old farmer, "Were they ALL dead?"

The old farmer replied, "Well, some of them said they weren't, but you know how them politicians lie."

Agency Test

The LAPD, the FBI, & the CIA are all trying to prove that they are the best at apprehending criminals. The President decides to test them. He releases a rabbit into a forest and each of them has to catch it.

The CIA goes in. They place animal informants throughout the forest. They question all plant and mineral witnesses. After three months of extensive investigations they conclude that the rabbit does not exist.

The FBI goes in. After two weeks with no leads they burn the forest, killing everything in it, including the rabbit and they make no apologies. The rabbit had it coming.

The LAPD goes in. They come out two hours later with a badly beaten bear. The bear is yelling: "Okay, okay, I'm a rabbit, I'm a rabbit."

911 Call

This little old lady calls 911. When the operator answers she yells, "Help, send the police to my house right away! There's a damn Democrat on my front porch and he's playing with himself."

"What?" the operator exclaimed. "I said there is a damn Democrat on my front porch playing with himself and he's weird; I don't know him and I'm afraid! Please send the police!" the little old lady repeated.

"Well, now, how do you know he's a Democrat?"

"Because, you damn fool, if it was a Republican, he'd be screwing somebody!"

Government Contractors

Three contractors were touring the White House on the same day. One was from New York, another from Missouri, and the third from Florida. At the end of the tour, the guard asked them what they did for a living.

When they each replied that they were contractors the guard said "Hey, we need one of the rear fences redone. Why don't you guys look at it and give me a bid." So to the back fence they went.

First up was the Florida contractor. He took out his tape measure and pencil, did some measuring and said, "Well I figure the job will run about $900. $400 for materials, $400 for my crew, and $100 profit for me."

Next was the Missouri contractor. He also took out his tape measure and pencil did some quick figuring and said, "Looks like I can do this job for $700. $300 for materials, $300 for my crew, and $100 profit for me."

Then the guard asks the New York contractor how much. Without so much as moving the contractor says $2700. The guard, incredulous looks at him says "You didn't even measure like the other guys! How did you come up with such a high figure?"

"Easy" says the contractor from New York, "$1,000 for me, $1,000 for you and we hire the guy from Missouri."

Brass Rat

A man walks into an antique store, and starts looking around. All of the sudden he spies a huge BRASS RAT in the corner. He falls in love with it, and so he takes it to the cashier.

"The rat, eh?" says the old grizzly cashier. "Um, yeah...how much?" replies our friend. "Well, five bucks for the rat—but 200 dollars for the story" he replied. "I'll just take the rat, without the story" says the customer.

He leaves the store, his precious brass rat tucked under his arm. Soon he begins to notice that a few rats are following him. He walks a few more blocks and the number of rats behind him increased . This continued, until there were virtually millions of rats behind him. Afraid of this mass following him, the man ran to the sea and threw the rat in. All of the rats plunged in after it, and met their watery deaths.

The man ran back to the antique store. The old cashier was chuckling to himself. "So now do you want the story?"

"No", said the man "but have you got any brass Democrats?'

Bill and Hillary

Quotes – Remember these?

"I'm not going to have some reporters pawing through our papers. We are the president."
- Hillary Clinton commenting on the release of subpoenaed documents

"I did not have sex with that woman... (long pause) Monica Lewinsky."
- President Bubba Clinton

"The president has kept all of the promises he intended to keep."
- Clinton lacky George Stephanopolous

Observations and Questions

> Bill Clinton has been testing a new product by the makers of the Clapper: Clap once and your pants drop to your knees. Clap twice and they return to their normal position.

> Bill's nickname for Hillary is, "My little buttercup" — His nickname for Monica? "My little suction cup"

> Did you hear that Clinton has broken the 11th. commandment? "Thou shall not place thy rod in thy staff"

> How do you get on Bill's good side? — Suck up!

> If Monica was a bird, what kind would she be? — A swallow.

> Monica Lewinsky didn't have a political bone in her body until she went to Washington D.C.

> What did Clinton say when Monica called him a creep? — "I'll thank you to keep a civil tongue on my head!"

> What does Clinton like more that roses on his piano? — Tulips on his organ.

> What is Kenneth Starr's opinion of his chances of prosecuting the President: — "Its an open and slut case!"

> What song did Clinton write for Monica? — "For she's a jolly good fellator"

> What's the difference between Monica Lewinsky and the rest of us? — When we want some dick in the White House, we just vote.

> Why can't Monica become a spy? — Because she spits everything out when the debriefing's done.

> Why did Bill stop playing the saxophone? — It was out of tune when Monica started playing his organ.

Presidential Parrot

The Clintons left the White House for a vacation, leaving a man behind to take care of the pets, including the Clintons' parrot. Unfortunately, two days into the vacation, the parrot died. The man was frantic; he began searching Washington DC pet shops for a replacement. Finally, in a small pet shop in a questionable neighborhood, he found a bird that was a perfect match. The shopkeeper, realizing where the bird was headed, advised against the purchase.

"This parrot spent years in one of the filthiest whorehouses in DC, and his vocabulary is just awful—not fit for the First Family."

"I don't care," said the man. He's a perfect match, and I've GOT to have him."

The purchase was made, and the replacement bird was installed on the White House perch.

In due course, the Clintons returned. As Hillary walked past, the bird chimed in, "Too old, too old!" As Chelsea walked past, the bird observed, "Too young, too young!"

As the President walked past, the parrot announced, "HELLOOO, Bill!!

First Pitch

Bill and Hillary are at the first baseball game of the season. The umpire walks up to the VIP section and says something.

Suddenly Clinton grabs Hillary by the collar and throws her over the wall onto the field.

The stunned umpire shouts, "No, Mr. President! I said, 'Throw the out the first PITCH!'"

Tragedy

Bill Clinton is visiting a school. In one class, he asks the students if anyone can give him an example of a "tragedy."

One little boy stands up and offers that "If my best friend who lives next door was playing in the street when a car came along and killed him, that would be a tragedy."

"No," Clinton says, "That would be an ACCIDENT."

A girl raises her hand. "If a school bus carrying fifty children drove off a cliff, killing everyone involved... that would be a tragedy."

"I'm afraid not," explains Clinton. "That is what we would call a GREAT LOSS."

The room is silent; none of the other children volunteer. "What?" asks Clinton, "Isn't there any one here who can give me an example of a tragedy?"

Finally, a boy in the back raises his hand. In a timid voice, he says: "If an airplane carrying Bill & Hillary Clinton were blown up by a bomb, *that* would be a tragedy."

"Wonderful!" Clinton beams. "Marvelous! And can you tell me WHY that would be a tragedy?"

"Well," says the boy, "because it wouldn't be an accident, and it certainly would be no great loss!"

Horses

A guy goes into the saloon in a little town in Montana. He has a few beers and then he says "Clinton is a horse's ass!" - and the guy standing next to him bashes him upside the head.

After he recovers from that and has a few more, he says "Clinton and his boss Hillary are both horses' asses!" Several people give him dirty looks, and the two nearest guys beat the shit out of him.

A few minutes later, he recovers, looks around the room and yells, "I still say Clinton is a horse's ass!!." Everybody in the place jumps him, and he is beaten to a pulp.

Hours later, he wakes up and everyone is gone except for the bartender. "Wow," he says, "I didn't know there were that many people left who were stupid enough to be democrats."

The bartender says, "Why, there wasn't a democrat in the house — they're all horse ranchers!"

Dinner

Bill and Hillary are at a restaurant. The waiter tells them tonight's specials are chicken almandine and fresh fish. "The chicken sounds good; I'll have that," Hillary says. The waiter nods. "And the vegetable?" he asks. "Oh, he'll have the fish," Hillary replies.

Politically Correct Ant & The Grasshopper

The Original Version:
The ant works hard in the withering heat all summer long, building his house and laying up supplies for the winter. The grasshopper thinks he is a fool and laughs and dances and plays the summer away.

Come winter, the ant is warm and well fed. The grasshopper has no food or shelter so he dies out in the cold.

The New Liberal Version:
It starts out the same, but when winter comes, the shivering grasshopper calls a press conference and demands to know why the ant should be allowed to be warm and well fed while others are cold and starving. CBS, NBC and ABC show up and provide pictures of the shivering

grasshopper next to film of the ant in his comfortable home with a table filled with food.

America is stunned by the sharp contrast. How can it be that, in a country of such wealth, this poor grasshopper is allowed to suffer so? Then a representative of the NAAGB (The National Association for the Advancement of Green Bugs) shows up on NightLine and charges the ant with "Green Bias" and makes the case that the grasshopper is the victim of 30 million years of greenism. Kermit the frog appears on Oprah with the grasshopper, and everybody cries when he sings "It's Not Easy Being Green."

Bill and Hillary Clinton make a special guest appearance on the CBS Evening News and tell a concerned Dan Rather that they will do everything they can for the grasshopper who has been denied the prosperity he deserves by those who benefited unfairly during the Reagan summers, or as Bill refers to it, the "Temperatures of the 80's." Richard Gephardt exclaims in an interview with Peter Jennings that the Ant has gotten rich off the "back of the grasshopper", and calls for an immediate tax hike on the Ant to make him pay his "fair share."

Finally the EEOC drafts the "Economic Equity and Anti-Greenism Act", RETROACTIVE to the beginning of the summer. The ant is fined for failing to hire a proportionate number of green bugs and, having nothing left to pay his retroactive taxes, his home is confiscated by the government. Hillary gets her old law firm to represent the grasshopper in a defamation suit against the ant, and the case is tried before a panel of federal judges that Bill appointed from a list of single-parent welfare moms who can only hear cases on Thursday afternoon between 1:30 and 3:00 PM when there are no talk shows scheduled.

The ant loses the case.

The story ends as we see the grasshopper finishing up the last bits of the ant's food while the government house he's in - which just happens to be the ant's old house - crumbles around him since he doesn't know how to maintain it. The ant has disappeared in the snow. And on the TV, which the grasshopper bought by selling most of the ant's food, they are showing Bill Clinton standing before a wildly applauding group of Democrats announcing that a new era of "Fairness" has dawned in America.

Q and A

Q. Bill and Hillary are on a sinking boat. Who gets saved?
A. The nation.

Q. What does Bill say to Hillary after having sex?
A: "Honey, I'll be home in 20 minutes."

Q. What do you get when you cross a crooked lawyer with a crooked politician?

A. Chelsea Clinton

Q: How many presidential candidates does it take to change a light bulb?
A: Fewer and fewer all the time.

Q: How many believable, competent, "just-right-for-the-job" presidential candidates does it take to change a light bulb?
A: It's going to be a dark 4 years, isn't it?

Bill and the Pope

Oddly enough, Bill Clinton and The Pope die on the same day. Being that Bill predeceased The Pope by a few minutes, and heaven being a place where rules are viewed as sacrosanct, Bill's application at the Pearly Gates was processed first. It took quite some time and he was finally admitted at 5 pm on the button.

The gate keeper politely informed the Pope that rules stipulated he could process no applications after 5 pm HST (Heaven Standard Time), but assured The Holy Father that his application would be expedited first thing the next morning.

Bright and early the next day, St. Peter himself is waiting for the Pope, quite embarrassed, as you may imagine. St. Peter apologizes profusely and says, "I can't apologize enough for your inconvenience. I'll handle your application myself and get you right in. In fact, I can promise you that, as soon as we get through here, I'll personally introduce you to Bill Clinton, how's that?"

The Holy Father said, "Thank you, but Bill Clinton is not the person I'm most looking forward to meeting."

"Who would you like to meet then?" inquired St. Peter. "The Virgin Mary." "Darn, you're one day too late."

Vacation

Last summer, the President and Mrs. Clinton were vacationing in their home state of Arkansas. On a venture on day, they stopped at a service station to fill up the car with gas. It seemed that the owner of the station was once Hillary's high school love.

They exchanged hellos, and went on their way.

As they were driving on to their destination, Bill put his arm around Hillary and said, "Well, honey, if you had stayed with him, you would be the wife of a service station owner today."

She smiled and replied, "No, if I had stayed with him, he would be President of the United States."

Heaven

Bill Clinton, Hillary Clinton and Al Gore are in an airplane flying across the country. The plane crashes and they all go to heaven. God comes down from his throne and looks the three of them over and says to Bill, "I'm God, who are you?"

Bill looks at him and says "I'm Bill Clinton, and I was the President of the United States!"

God thinks for a minute and says, "Hmmmm, that's a pretty important job—you come sit here on my right."

Then God looks at Al Gore. "I'm God, who are you?"

And Al says, "I'm Al Gore, and I was the Vice President of the United States!"

God thinks for a minute and says "Hmmmm, that's a pretty important job—you come sit here on my left."

Then God looks at Hillary. "I'm God, who are you?"

"I'm Hillary Rodham Clinton—and I believe you're in MY Chair!"

Green Side Up

A woman calls a general contractor to her home to discuss the interior painting she wants done. In the living room, she says "I'd like this room to be done in a soft blue." The contractor nods, writes this down, and then sticks his head out the window and shouts "Green side up!"

The woman thinks this is odd, but says nothing. In the next room, she says "I think off white would be perfect for this room." The contractor nods, writes down the information, then sticks his head out the window and shouts "Green side up!"

The woman is concerned, but keeps her silence. In another room, she says "The should be painted in a warm color." The man nods, writes as he mutters "warm color," and then again sticks his head out of the window and outs "Green side up!"

The woman can take it no more. "I don't understand," she says. "No matter what color I tell you, you always shout 'green side up!' What is going on?"

"I'm sorry," the contractor says, "But I've got a crew of Clinton supporters across the street laying sod."

Bill's Driver

Bill Clinton and his driver were cruising along a country road one night when all of a sudden they hit a pig, killing it instantly. Bill told his driver to go up to the farm house and explain to the owners what had happened.

About 1 hour later Bill sees his driver staggering back to the car with a bottle of wine in one hand, a cigar in the other and his clothes all ripped and torn.

"What happened to you," asked Bill.

Well, the Farmer gave me the wine, his wife gave me the Cigar and his 19 year old daughter made mad passionate love to me.

"My God, what did you tell them," asks Clinton.

The driver replies, "I'm Bill Clinton's driver, and I just killed the pig."

Message

Clinton is looking out of the window and he notices that someone has spelled out the message, "BILL SUCKS!" in urine on the White House Lawn. Furious, he orders the FBI to take urine and handwriting samples from every member of the White House staff and find the culprit immediately.

A week later, the FBI director calls. "Mr. President, I have good news and bad news," he says. "The good news is that the urine belongs to Vice President Gore."

"And the bad news?" Clinton demands.

"Well, sir, the handwriting belongs to Hillary!"

Things Overheard Coming From The Oval Office

1. Are you sure it's in?
2. Are you sure that Al Gore started this way?
3. I knew that a lot of things come across your desk, I just never thought that I would be one of them.
4. I thought that all of those notches in your desk were from Sox sharpening his claws.
5. If I could convince Hillary to do that just once...
6. If this doesn't leak out, I'll be ruined.
7. If this leaks out, I'll be ruined.
8. If you think that's 8 inches, I can see why you thought your last budget was balanced.
9. I've always said, "I want to be a 'hands-on' president."
10. Maybe Chelsea can hook me up with some of those sorority babes.
11. Now you know why they call me 'Slick Willie'.
12. Somehow, I don't think that Alan Greenspan would explain inflation that way.
13. What do you mean "falsie inspection?" I don't remember a no falsies clause in my contract.
14. When you asked me to look at the presidential pole, I thought you meant the latest Gallup Survey.
15. When you said that you had your finger on the pulse of the nation, this isn't exactly what I thought you meant.
16. You are a White House intern; Well, now it's your turn.

Razor Backs

Clinton returns from a vacation in Arkansas and walks down the steps of Air Force One with two pigs under his arms. At the bottom of the steps, he says to the honor guardsman, "These are genuine Arkansas Razor-Back Hogs. I got this one for Chelsea and this one for Hillary."

The guardsman replies, "Nice trade, Sir."

Jogging

Bill Clinton went jogging one morning and came upon the Washington monument. He said, "George, what should I do?" After a few seconds George replied, "Abolish the IRS and start over."

Bill thought about this for a few seconds and continued jogging. Shortly he came upon the Jefferson Memorial and stopped. He said "Tom, what should I do?"

After a few seconds Tom replied, "Abolish welfare and start over." Bill continued jogging after thinking about this and came upon the Lincoln Memorial. He said, "Abe, what should I do?"

After a few seconds Abe replied "Why don't you take the night off and go to the theater?"

Airplane Trouble

An airplane was about to crash and there were 5 passengers left, but only 4 parachutes!

The first passenger, George W. Bush said, "I am the President of the United States, and I have a great responsibility, being the leader of nearly 300 million people, and a superpower, etc., and I am also the smartest president ever." So he takes the first parachute, and jumps out of the plane.

The second passenger said," I'm Rasheed Wallace, one of the best basketball players in the NBA, and the Portland Trailblazers need me, so I cannot afford to die." So he takes the second parachute, and leaves the plane.

The third passenger, Hillary Clinton, said; "I am the wife of the former President of the United States, I am New York's Senator, and I am the smartest woman in the world." She takes the third parachute and exits the plane.

The fourth passenger, an old man, says to the fifth passenger, a 10 year old Boy Scout, "I am old and frail and I don't have many years left, so as a Christian gesture and a good deed, I will sacrifice my life and let you have the last parachute."

The Boy Scout said, It's okay, there's a parachute left for you. The world's smartest woman took my backpack."

Rednecks

Personal Hygiene

Unlike clothes and shoes, a toothbrush should never be a hand-me-down item. While ears need to be cleaned regularly, this is a job that should be done in private using one's OWN truck keys. Proper use of toiletries can forestall bathing for several days. However, if you live alone, deodorant is a waste of good money. Dirt and grease under the fingernails is a social no-no, as they tend to detract from a woman's jewelry and alter the tastes of finger foods. Plucking unwanted nose hair is time-consuming work. A cigarette lighter and a small tolerance for pain can accomplish the same goal and save hours. Its a good idea to keep a bucket of water handy when using this method.

Dining Out

When decanting wine, make sure that you tilt the paper cup and pour slowly so as not to "bruise" the fruit of the vine. If drinking directly from the bottle, always hold it with your fingers covering the label. Remember to leave a generous tip for good service. After all, their mobile home costs just as much as yours.

You May Be A Redneck If ... # 1

- ➤ After the Prom, you drove the truck while your date hit road signs with beer bottles.
- ➤ All of your 4 letter words are two syllables.
- ➤ Birds are attached to your beard.
- ➤ Every socket in your house breaks the fire code.
- ➤ Have a Hefty Bag for a passenger side window.
- ➤ Hitchhikers won't get into the car with you.
- ➤ MOTEL 6 turns off the lights when they see you coming.
- ➤ People hear your car is shaped like a bare foot.
- ➤ The dog cannot watch you eat without gagging.
- ➤ The Dog Catcher calls for a backup unit when visiting your house.
- ➤ The Home Shopping Operator recognizes your voice.
- ➤ The taillight covers of your car are made of red tape.
- ➤ There are more than five McDonald's bags currently in the floorboard of your car.
- ➤ There has ever been crime-scene tape on your bathroom door.
- ➤ You actually know which type of leaf is a good sub for toilet pap.
- ➤ You bought a VCR because wrestling is on while you are at work.

> You burn your yard rather than mow it.
> You call your boss "dude."
> You can spit without opening your mouth.
> You clean your fingernails with a stick.
> You consider "Outdoor Life" deep reading.

Entertaining In Your Home

A centerpiece for the table should never be anything prepared by a taxidermist. Do not allow the dog to eat at the table ... no matter how good his manners are. Be considerate of your guests. Point out in advance where the injury- threatening springs are located on the sofa. If your dog falls in love with a guest's leg, have the decency to leave them alone for a few minutes.

Dating (Outside The Family)

Always offer to bait your date's hook, especially on the first date. No matter how broke you are, never take your date flowers that were stolen from a cemetery . Be aggressive. Let her know you are interested: "I've been wanting to go out with you since I read that stuff on the men's bathroom wall two years ago." Establish with her parents what time she is expected back. Some will say 10:00. Others might say "Monday." If the latter is the answer, it's the boy's responsibility to get her to school on time. If a girl's name does not appear regularly on a bathroom wall, water tower, or an overpass, odds are good that the date will end in frustration. Even if you cannot get a date, avoid kidnapping. It's bad for your reputation.

You May Be A Redneck If ... # 2

> You consider your license plate personalized because your father made it.
> You cut your toenails in front of company.
> You discover a Volvo IS a part of a woman's anatomy, and you try to test drive it.
> You dog has a litter of puppies on the living room floor and nobody notices.
> You ever hit on someone in a V.D. Clinic.
> You go to a stock race and you don't even need a program.
> You hammer bottle caps into the frame of your front door to make it look nice.
> You have a rag for a gas cap.
> You have every episode of "Hee-Haw" on tape.
> You have the local taxidermist's number on speed dial.
> You keep a can of RAID on the kitchen table.
> You own a denim leisure suit.

> ➤ You own a homemade fur coat.
> ➤ You own more than 3 shirts with the sleeves cut off.
> ➤ You pick your teeth from a catalog.
> ➤ You prefer car keys to Q-tips.
> ➤ You read the Auto Trader with a highlight pen.
> ➤ You refer to the time you won a free case of motor oil as "the day my ship came in."
> ➤ You see no need to stop at rest stops because you have an empty milk jug in the car.
> ➤ You show your boyfriend you really love him by carving his name into your arm.
> ➤ You take a fishing pole into Sea World.

Theater Etiquette

Crying babies should be taken to the lobby and picked up immediately after the movie has ended. Refrain from talking to characters on the screen. Tests have proven they cannot hear you.

Weddings

Livestock usually is a poor choice for a wedding gift. Is it okay to bring a date to a wedding? Not if you are the groom. When dancing, never remove undergarments, no matter how hot it is. Kissing the bride for more than 5 seconds may get you cut. A bridal veil made of window screen is not only cost effective but also a proven fly deterrent. For the groom, at least rent a tux. A leisure suit with a cummerbund and a clean bowling shirt can create a natty appearance. Though uncomfortable, say yes to socks and shoes for this special occasion.

Driving Etiquette

Dim your headlights for approaching vehicles, even if the gun is loaded and the deer is in sight. When approaching a four-way stop, the vehicle with the largest tires always has the right of way. Never tow another car using pantyhose and duct tape. When sending your wife down the road with a gas can, it is impolite to ask her to bring back beer. Never relieve yourself from a moving vehicle, especially when driving. Do not remove the seats from the car so that all your kids can fit in. Do not lay rubber while traveling in a funeral procession.

You May Be A Redneck If ... # 3

> ➤ You think "taking out the trash" means taking your in-laws to a movie.
> ➤ You think a chain saw is a musical instrument.

- You think a subdivision is part of a math problem.
- You think a turtle-neck is an ingredient for soup.
- You think a Volvo is a part of a woman's anatomy.
- You think Campho-Phenique is a miracle drug.
- You think Dom Perignon is a Mafia leader.
- You think the French Riviera is a foreign car.
- You think the stock market has a fence around it.
- You view the upcoming family reunion as a chance to meet men.
- You wonder how service stations keep their restroom so clean.
- Your brother in-law is also your uncle.
- Your CB antenna is a danger to low-flying airplanes.
- Your entire family has ever sat around waiting for a call from the Governor to spare a loved one.
- Your father executes the "Pull my finger" trick during Christmas dinner.
- Your grandmother has ever been asked to leave the bingo hall because of her language.
- Your house doesn't have curtains but your truck does.
- Your kids take a siphon hose to "Show and Tell."
- Your lover has "ammo" on his Christmas list.
- Your mother keeps a spit cup on the ironing board.
- Your mother does not remove the Marlboro from her mouth before telling the state trooper to kiss her ass.

Tips For All Occasions

Never take a beer to a job interview or ask if they press charges. Always identify people in your yard before shooting at them. Always say "Excuse me" after getting sick in someone else's car. It's considered tacky to take a cooler to church. Even if you're certain that you are included in the will, it's considered tacky to drive a U-Haul to the funeral home. The socially refined never fish coins out of public toilets, especially if other people are around. If you have to vacuum the bed, it's time to change the sheets. Always provide an alibi to the police for family members.

Honeymoon

A redneck and his new bride were on their honeymoon. The husband jumps into bed to wait for his wife to get herself ready. The wife comes out of the bathroom in a sexy negligee and says, "Honey, I have something to tell you. I'm a virgin."

The man grabs his clothes and rushes out of the house yelling at the top of his lungs. He heads straight to his father's house. When he gets there his father says, "Son, what are you doing here? You're supposed to be on your honeymoon."

The son says, "Dad, my new wife told me a big secret of hers. She's a virgin."

"Damn son. You did the right thing by leaving. If she wasn't good enough for her kinfolk, she sure as hell isn't good enough for ours!"

You May Be A Redneck If ... # 4

- ➤ Your pocket knife doubles as a tooth pick.
- ➤ Your Significant Other has ever said, "Come move this transmission so I can take a bath!"
- ➤ Your stereo speakers used to belong to the Moonlight drive-in Theater.
- ➤ Your toilet paper has page numbers on it.
- ➤ Your wife has a beer belly and you find it attractive.
- ➤ You're considered an expert on worm beds.
- ➤ You've ever bathed with flea and tick soap.
- ➤ You've ever been involved in a custody fight over a hunting dog.
- ➤ You've ever been kicked out of the KKK for being a BIGOT.
- ➤ You've ever been too drunk to fish.
- ➤ You've ever bought a used hat.
- ➤ You've ever financed a tattoo.
- ➤ You've ever given rat traps as gifts.
- ➤ You've ever had to scratch your sisters name out of a message that begins," For a good time, call..."
- ➤ You've ever heard a sheep bleat and had romantic thoughts.
- ➤ You've ever hit a deer with your car...deliberately.
- ➤ You've ever raked leaves in your kitchen.
- ➤ You've ever stolen toilet paper.
- ➤ You've ever stood in line to have your picture taken with a freak of nature.
- ➤ You've ever worn a tube-top to a wedding.
- ➤ You've totaled every car you've ever owned.

Blondes

Brunette, Blonde and a Red-Head # 1

Three girls are running from the cops on a farm. There's a brunette, blonde and a red-head. They come to a barn and decide to hide in there to elude the police. When they get into the barn they notice three sacks and decide each one will hide in a sack. In the event that the cops do come in, they would just make some barn noise.

So, all three are waiting and the police go into the barn and check it out. One officer sees the sacks and goes over to check it out.

He goes to the one with the brunette in it and gives it a little kick. "Meow", says the brunette. "Aw it's just a cat." says the cop.

He then goes over to the sack with the red-head in it and gives it a kick. "Woof", she says. "Aw, it's just a dog," says the cop.

He then goes over to the sack with the blonde in it and gives it a kick. And the blonde says "Potatoes."

Blonde's Revenge

Q. What's black and blue and brown and laying in a ditch?
A. A brunette who's told too many blonde jokes.

Q. What do you call going on a blind date with a brunette?
A. Brown-bagging it.

Q. What's the real reason a brunette keeps her figure?
A. No one else wants it.

Q. What do you call a brunette in a room full of blondes?
A. Invisible.

Q. What's a brunette's mating call?
A. "Has the blonde left yet?"

Q. Why don't brunettes make good cattle ranchers?
A. Because they cannot keep their calves together.

Q. What goes screech-vroom, screech-vroom?
A. That's a brunette driving through a flashing red light.

Blonde Easter

Three blondes died and are at the pearly gates of heaven. St. Peter tells them that they can enter the gates if they can answer one simple question. St. Peter asks the first blonde, "What is Easter?"

The first blonde replies, "Oh, that's easy! It's the holiday in November when everyone gets together, eats turkey, and are thankful..."

"Wrong!", replies St. Peter, and proceeds to ask the second blonde the same question, "What is Easter?"

The second blonde replies, "Easter is the holiday in December when we put up a nice tree, exchange presents, and celebrate the birth of Jesus."

St. Peter looks at the second blonde, shakes his head in disgust, tells her she's wrong, and then peers over his glasses at the third blonde and asks, "What is Easter?"

The third blonde smiles confidently and looks St. Peter in the eyes, "I know what Easter is."

"Oh?" says St. Peter, incredulously.

"Easter is the Christian holiday that coincides with the Jewish celebration of Passover. Jesus and his disciples were eating at the last supper and Jesus was later deceived and turned over to the Romans by one of his disciples. The Romans took him to be crucified and he was stabbed in the side, made to wear a crown of thorns, and was hung on a cross with nails through his hands. He was then buried in a nearby cave which was sealed off by a large boulder."

St. Peter smiles broadly with delight.

The third blonde continues, "Every year the boulder is moved aside so that Jesus can come out. If he sees his shadow, there will be six more weeks of winter."

Brunette, Blonde and a Red-Head # 2

A blonde, brunette and a red-head were all at a bar one night. The owner was telling them about the mirror in the bathroom. If they go and say something true in mirror they will get it back times three. But if they said something false the mirror would swallow them.

The brunette goes in first. She says: I'm beautiful. She walked out looking 3X better.

The red-head was next. She said: I'm smart. She came out 3X smarter.

The blonde goes in last. Walks up to the mirror and says: Well, I think... and Poof! The mirror swallows her.

Elevator Ride

A blonde & brunette are in an elevator. On the third floor a man gets on who is perfect;

84

3-piece suit, great build with a nice butt. Unfortunately, they both noticed, he had really bad dandruff.

The man got off on the 5th floor. Once the doors closed the brunette turned to the blonde and said, "Someone should give him 'Head & Shoulders.'"

To which the blonde replied, "How do you give 'Shoulders'?"

Machines

One day a blonde walks up to a Coke machine and puts in a coin. Out pops a coke. The blonde looks amazed and runs away to get some more coins. She returns and starts feeding the machine madly. A few minutes later a thirsty little kid walks up behind the blonde and watches her for a few minutes before stopping her and asking if he could get a soda.

The blonde spins around and shouts in his face, "Can't you see I'm winning?"

Driving

A blonde flips her car on a major highway. A few minutes later a officer shows up and asks the blond what happened.

The blond says, "there was a tree in the middle of the road, so I swerved to the right and there was another tree, so I swerved to the left and there was another tree, so I swerved back to the to the right and there was another tree, SO! I swerved back to the left and there was another tree."

The officer said, "ma'me, there isn't a tree around here for thirty miles, that was your air freshener!"

Counting

There was a brunette standing along side a busy road chanting "88, 88, 88, 88..." until a blonde came up to her and said, "that looks like fun, can I try?"

The brunette said sure so the blonde chanted, "88, 88, 88, 88.." "Well," said the brunette, "that is fun. But what is even more fun is if you say it in the middle of the street."

So the blonde said "OK" and stood in the middle of the street. "88, 88, 88, 88-" BAM! she was run over by a car, completely flattened.

Along the side of the road, the brunette began to chant, "89, 89, 89, 89...

House Fire

This blonde calls up the fire department and says, "My house is on fire! Come quick!"

The fire chief asks, "How do we get there?"

The blonde replies, "Don't you still have those big red trucks?"

Swimming the Channel

A blonde woman competed with a brunette woman and a redheaded woman in the Breast Stroke division of an English Channel swim competition. The brunette came in first, the redhead second. The blonde woman finally reached shore completely exhausted.

After being revived with blankets and coffee, she remarked, "I don't want to complain, but I think those other two girls used their arms."

Pilots

An Air Traffic Controller heard a "mayday" call from a small airplane in distress being flown by a blonde. The controller asked; "aircraft calling mayday; what is your height and your position?"

The blonde replied; "Well, I am 5'6" tall and I am sitting in the cockpit."

Shorts

> Do you know why you always see blondes in groups of 18 at movie theaters? Because the sign often says "under 17 not admitted."
> How do you get a blonde to commit suicide? Put a mirror at the bottom of the swimming pool.
> How many blondes does it take to make chocolate-chip cookies ... 10 ... one to mix the dough and nine to peel the M & M's.
> The blonde painted an X on the bottom of the rented boat so she could find the same fishing spot again ... and her blonde friend called her an idiot because she knew that they might not get the same boat next time.
> What did the blonde say when she looked into a box of Cheerios ... oh look, donut seeds.
> What do you call 20 blondes in a freezer ... frosted flakes.
> What do you call a blonde in a tree with a brief-case ... branch manager.
> What do you call a brunette with a blonde on both sides ... an interpreter.
> What do you call a smart blonde ... a golden retriever.
> What do you call four blondes in a Volkswagen ... far-from-thinkin.
> What do you do when a blonde throws a pin at you? Run like Hell...she's got a hand grenade in her mouth.
> What does a blonde owl say ... what, what ...
> What is it called when a blonde blows in another blonde's ear ... data transfer.

> Why can't blondes put in light bulbs ... they keep breaking them with the hammer.
> Why did the blonde have blisters on her lips? From trying to blow out light bulbs.
> Why did the blonde keep a coat hanger in her back seat ... in case she locks the keys in her car.
> Why do blondes have more fun ... they are easier to keep amused.
> Why were blondes created ... because dogs cannot bring beer from the fridge ... why were brunettes created ... neither could the blondes.

Walkman

A blonde goes into the hair parlor with her walkman on her head.

"I need to take the walkman off", says the stylist. "You can't I'll die", the blonde replies.

But I can't cut your hair with the walkman on your ears." "You can't take it off, I'll die." the blonde again replies.

Flustered the hair stylist grabs the walkman and takes it off of the head of the blonde and the blonde dies!

The police come and listen to the walkman - it is repeating "breath in", breath out, breath in..."

Medical

The Short History of Medicine

2000 B.C. - Here, eat this root.
1000 A.D. - That root is heathen. Here, say this prayer.
1850 A.D. - That prayer is superstition. Here, drink this potion.
1940 A.D. - That potion is snake oil. Here, swallow this pill.
1985 A.D. - That pill is ineffective. Here, take this antibiotic.
2000 A.D. - That antibiotic is artificial. Here, eat this root.

Sperm Donor #1

A handsome young man walks into a sperm bank and declares I'm of royal blood and an I.Q. of 165, I'd like to make a donation. The nurse gives him a sealed cup and directs him to a private room. 20 minutes later the man hasn't come out, the nurse knocks on the door.

"Is there a problem?" she asked.

"I'm so embarrassed", he replied. "I used my right hand. I used my left hand. I poured cold water on it and hot water on it. Could you help me?"

The nurse replied "I don't usually do this but you are kind of cute..." She gets on her knees and begins to blow him.

"I really appreciate this, but I need help getting the cap off the jar!"

Seminal Buildup Disorder

A medical student specializing in sexual disorders, makes arrangements to visit a sexual disorder clinic. The chief doctor is showing him around, discussing the cases and the facilities, when the student sees a patient masturbating right there in the hallway.

"What condition does he have?" asks the student.

"He suffers from Seminal Buildup Disorder," the doctor replies. "If he doesn't obtain sexual release forty to fifty times a day, he'll go into a coma."

The student takes some notes on that and they continue down the hall. As they turn the corner, he sees another patient with his pants around his ankles, receiving oral sex from a beautiful nurse.

"What about him?" the student asks. "What's his story?"

"Oh, it's the same condition," the doctor replies. "He just has a better health plan."

Doctor's Visit

A beautiful woman walks into a doctors office and he is awestruck. All his professionalism goes out the window. He tells her to take off her pants and he starts rubbing her thighs.

He says "Do you know what I am doing?" She replies "Yes, checking for abnormalities."

He tells her to take off her shirt and bra and he starts rubbing her breasts. He says "Do you know what I am doing now?" She replies "Yes, checking for lumps and cancer."

Finally, he tells he takes off her panties, lays her on the table, gets on top of her, and starts having sex with her. He says "Do you know what I am doing now?" She replies "Yes, getting harpies. That's why I am here."

Sleep Better

An elderly woman went into the doctor's office. When the doctor asked why she was there, she replied, "I'd like to have some birth control pills."

Taken aback, the doctor thought for a minute and then said, "Excuse me, Mrs. Santos, but you're 75 years old. What possible use could you have for birth control pills?"

The woman responded, "They help me sleep better."

The doctor thought some more and continued, "How in the world do birth control pills help you to sleep?"

The woman said, "I put them in my granddaughter's orange juice and I sleep better at night."

Farmer in the Pharmacy

A farmer walked into a drug store and said to the pharmacist, "I want me one of them thar condoms with pesticides on it. Where do I find 'em?"

The pharmacist replied, "Oh sir, you must mean that you want the condoms with SPERMICIDE, not pesticide. They're on aisle 4."

"No, no, I want me them thar condoms with PESTICIDE on it," growled the farmer.

"Sir," said the pharmacist, exasperated from explaining, "PESTICIDE is for killing insects, SPERMICIDE is for killing sperm. I'm sure that you mean spermicide instead of pesticide."

"Listen here," argued the farmer, "my wife's got a bug up her ass and I'm a goin' huntin' for it. Like I said, I want me one of them condoms with PESTICIDE on it!"

Things You DON'T Want To Hear While On The Operating Room Table

- ➢ Accept this sacrifice, O Great Lord of Darkness.
- ➢ Anyone see where I left that scalpel?
- ➢ Bo! Bo! Comeback with that! Bad Dog!
- ➢ And now we remove the subject's brain and place it in the body of the ape.
- ➢ Better save that. We'll need it for the autopsy.
- ➢ Could you stop that thing from beating; it's throwing

- ➢ Damn! Page 47 of the manual is missing.
- ➢ Don't worry. I think it is sharp enough.
- ➢ FIRE! FIRE! Everyone get out!
- ➢ Hell, the guy's got two of 'em.
- ➢ I wish I hadn't forgotten my glasses.
- ➢ Oh no! I just lost my Rolex.

- ➢ Oops! Hey, has anyone ever survived 500ml of this stuff before?
- ➢ Someone call the janitor - we're going to need a mop.
- ➢ That's cool! Now can you make his leg twitch?

- ➢ Wait a minute, if this is his spleen, then what's that?
- ➢ What do you mean "You want a divorce?"
- ➢ What's this doing here?

- ➢ my concentration off
- ➢ Damn, there go the lights again...
- ➢ Everybody stand back! I lost my contact lens!
- ➢ Hand me that...uh...that uh....thingie
- ➢ I hate it when they're missing stuff in here.
- ➢ Nurse, did this patient sign the organ donation card?
- ➢ OK, now take a picture from this angle. This is truly a freak of nature.
- ➢ She's gonna blow! Everyone take cover!!!

- ➢ Sterile, schmerile. The floor's clean, right?
- ➢ This patient has already had some kids, am I correct?
- ➢ Well folks, this will be an experiment for all of us.
- ➢ What do you mean he wasn't in for a sex change!
- ➢ Ya know, there's big money in kidneys.

Terminal Patient

A man goes to his doctor for a check up and the doctor says "I don't know how to tell you this, but you're going to die, and you only have six months left."

When the poor bloke gets home, he tells his wife he has AIDS and only has six months to live and goes out for a beer. He gets wired up and tells all his mates he has AIDS and only six months left.

Two days later he meets his doctor in the street and the doc says "I see you've come to terms with your terminal condition, everyone in town is talking about it. But tell me why are you telling everyone you have AIDS when I told you it was an inoperable brain tumor that's killing you?"

"Oh," said the man "I've come to terms with dying, but I don't want anyone screwing my wife after I'm gone!

Sperm Donor #2

An elderly man went to his doctor's office to get a sperm count. The doctor gave the man a jar and said, "Take this jar home and bring me back a sample tomorrow."

The next day, the man reappears at the doctor's office and gives him the jar, which is as clean and empty as on the previous day. The doctor asks what happened.

"Well, doc, it's like this. First I tried with my right hand, but nothing. Then I tried with my left hand, but nothing. Then I asked my wife for help. She tried with her right hand, but nothing. Then her left, but nothing. She even tried with her mouth, first with the teeth in, then with the teeth out, and still nothing. Hell, we even called up the lady next door, and she tried with both hands and her mouth too, but nothing."

The doctor was shocked. "You asked your NEIGHBOUR?"

The man replied, "Yep, but no matter what we tried, we couldn't get the damn jar open!"

Orange Problem

Guy goes to a doctor and says, "Doc, you've got to help me. My penis is orange." Doctor pauses to think and asks the guy to drop his pants so he can check. Damned if the guy's penis isn't orange.

Doc tells the guy, "This is very strange. Sometimes things like this are caused by a lot of stress in a person's life." Probing as to the causes of possible stress, the doc asks the guy, "How are things going at work?"

The guy responds that he was fired about six weeks ago. The doctor tells him that this must be the cause of the stress.

Guy responds, "No. The boss was a real asshole, I had to work 20-30 hours of overtime every week and I had no say in anything that was happening. I found a new job a couple of weeks ago where I can set my own hours, I'm getting paid double what I got on the old job and the boss is a really great guy." So the doc figures this isn't the reason.

He asks the guy, "How's your home life?"

The guy says, "Well, I got divorced about eight months ago." The doc figures that this has got to be the reason for all of the guys stress.

Guy says, "No. For years, all I listened to was nag, nag, nag. God, am I glad to be rid of that old bitch." So the doc takes a few minutes to think a little longer.

He inquires, "Do you have any hobbies or a social life?" The guy replies, "No, not really. Most nights I sit home, watch some porno flicks and munch on Cheetos."

Joseph J. Zajac III

Office Charge

A couple, aged 67, went to the doctor's office. The Doctor asked, "What can I do for you?"

The man said, "Will you watch us have sexual intercourse?"

The doctor looked puzzled but agreed. When the couple had finished, the doctor said, "There is nothing wrong with the way you have intercourse." And he charged them $20.00.

This happened several weeks in a row. The couple would make an appointment, have intercourse, pay the doctor and leave. Finally the doctor asked, "Just exactly what are you trying to find out?"

The old man said, "We're not trying to find out anything. She is married and we cannot go to her house. I am married and we cannot go to my house. Holiday Inn charges $32.00. Hilton Hotel charges $37.00. We do it here for $20.00 and I get $18.00 back from Medicare for a visit to the doctor's office."

Strange Dreams

A man goes to a doctor, complaining of strange dreams.

"Every night" he tells the doctor, "I dream of either a teepee or a wigwam. Over and over again, I see this teepee and this wigwam, and it's always the same teepee and wigwam."

"Ahhhhhhh," says the doctor, "the problem is obvious." "You're two tents."

Nurses

A handsome young lad went into the hospital for some minor surgery and the day after the procedure, a friend stopped by to see how the guy was doing.

The friend was amazed at the number of Nurses who entered the room in short intervals with refreshments, offers to fluff his pillows, make the bed, give back rubs, etc. "Why all the attention?" the friend asked. "You look fine to me."

"I know!" grinned the patient. "But the Nurses kinda formed a little fan club when they all heard that my circumcision required twenty-seven stitches."

Visit to the Vet

Women walks into the a vets office with an obviously dead dog in her arms. She is frantic while talking to the vet. "You've got to do something", she exclaims.

The vet tries to explain to her that the dog has passed on and that there is really nothing that he can do. After the woman begs and pleads he finally agrees to run some tests to verify that the dog is truly dead.

He puts the stiff dog on the table and grabs a nearby feline by the scruff of it's neck. The whole time the cat is hissing and clawing at the supposedly dead dog. He moves the cat over the dogs head. Nothing. It's chest. Nothing. It's stomach. Nothing. And finally, the dog's private parts. Nothing.

He then approaches the woman and tells her that yes, the dog is dead and hands her a bill for $250.00. The woman is enraged! "How can you justify $250.00 to examine my dead dog!" she exclaims.

"Simple", the vet replies, "$50.00 for the office visit and $200.00 for the cat scan."

Dentist

There's the woman who goes to the dentist. As he leans over to begin working on her, she grabs his balls. The dentist says, "Madam, I believe you've got a hold of my privates."

"Yes," the woman replies, "we're going to be careful not to hurt each other, aren't we?"

Sleeping with Your Patient

James and Nick were sitting in a pub, having a beer after work one day when Nick turns to James and says, "James you look troubled. What's the matter?"

Nick replies, "My conscience is bothering me. I slept with one of my patients today."

To which James replied, "Come on, Nick, don't let it bother you. You are not the first doctor to sleep with one of his patients.

And Nick replies, "True, very true, but damn it, James, I'm a veterinarian!"

Questions Answered on Health Care

Q. What does HMO stand for?

A. This is actually a variation of the phrase, "Hey, Moe!" Its roots go back to a concept pioneered by Doctor Moe Howard, who discovered that a patient could be made to forget about the pain in his foot if he was poked hard enough in the eyes. Modern practice replaces the physical finger poke with hi-tech equivalents such as voice mail and referral slips, but the result remains the same.

Q. Do all diagnostic procedures require pre-certification?

A. No. Only those you need.

Q. I just joined a new HMO. How difficult will it be to choose the doctor I want?

A. Just slightly more difficult than choosing your parents. Your insurer will provide you with a book listing all the doctors who were participating in the plan at the time the information was gathered. These doctors basically fall into two categories – those who are no longer accepting new patients, and those who will see you but are no longer part of the plan. But don't worry — the remaining doctor who is still in the plan and accepting new patients has an office just a half day's drive away!

Q. What are pre-existing conditions?

A. This is a phrase used by the grammatically challenged when they want to talk about existing conditions. Unfortunately, we appear to be pre-stuck with it.

Q. Well, can I get coverage for my pre-existing conditions?

A. Certainly, as long as they don't require any treatment.

Q. What happens if I want to try alternative forms of medicine?

A. You'll need to find alternative forms of payment.

Q. My pharmacy plan only covers generic drugs, but I need the name brand. I tried the generic medication, but it gave me a stomach ache. What should I do?

A. Poke yourself in the eye.

Q. I have an 80/20 plan with a $200 deductible and a $2,000 yearly cap. My insurer reimbursed the doctor for my out-patient surgery, but I'd already paid my bill. What should I do?

A. You have two choices. Your doctor can sign the reimbursement check over to you, or you can ask him to invest the money for you in one of those great offers that only doctors and dentists hear about, like windmill farms or frog hatcheries.

Q. What should I do if I get sick while traveling?

A. Try sitting in a different part of the bus.

Q. No, I mean what if I'm away from home and I get sick?

A. You really shouldn't do that. You'll have a hard time seeing your primary care physician. It's best to wait until you return, and then get sick.

Q. I think I need to see a specialist, but my doctor insists he can handle my problem. Can a general practitioner really perform a heart transplant right in his office?

A. Hard to say, but considering that all you're risking is the $10 co-payment, there's no harm giving him a shot at it.

Q. What accounts for the largest portion of health care costs?
A. Doctors trying to recoup their investment losses.

Q. Will health care be any different in the next century?
A. No, but if you call right now, you might get an appointment by then.

Men

Problems

A man jumps from an airplane and when he pulls his parachute cord it breaks. As he's plunging to his death, he sees a man rising rapidly into the air. As they cross paths, one falling toward the earth and the other rising away from it, the skydiver yells, "Excuse me! You wouldn't happen to know anything about parachutes would you?"

"Sorry, I don't." The other man yells back. "would you know anything about lighting gas stoves?"

Drunk Driving

One night, a police officer was staking out a particularly rowdy bar for possible violations of the driving-under-the-influence laws.

At closing time, he saw a fellow stumble out of the bar, trip on the curb, and try his keys on five different cars before he found his. Then, sat in the front seat fumbling around with his keys for several more minutes. During this time, everyone left the bar and drove off. Finally, the fellow started his engine and began to pull away.

The police officer was waiting for him. He stopped the driver, read him his rights, and administered the Breathalyzer test. The results showed a reading of 0.0. The puzzled officer demanded to know how that could be.

The driver replied, "Tonight, I'm the designated decoy."

Drink and Gamble

A bum asks a man for $2. The man asked, "Will you buy booze?"
The bum said, "No."
The man asked, "Will you gamble it away?"
The bum said, "No."
Then the man asked, "Will you come home with me so my wife can see what happens to a man who doesn't drink or gamble?"

Dog Story

Four men were bragging about how smart their dogs are. The first man was an Engineer, the second man was an Accountant, the third man was a Chemist, the fourth was a Government Worker. To show off, the Engineer called to his dog. "T-square, do your stuff." T-square trotted over to a desk, took out some paper and a pen and promptly drew a circle, a square, and a triangle. Everyone agreed that was pretty smart.

But the Accountant said his dog could do better. He called his dog and said, "Spreadsheet, do your stuff." Spreadsheet went out into the kitchen

96

and returned with a dozen cookies. He divided them into 4 equal piles of 3 cookies each. Everyone agreed that was good.

But the Chemist said his dog could do better. He called his dog and said, "Measure, do your stuff." Measure got up, walked over to the fridge, took out a quart of milk, got a 10 ounce glass from the cupboard and poured exactly 8 ounces without spilling a drop. Everyone agreed that was good.

Then the three men turned to the Government Worker and said, "What can your dog do?"

The Government Worker called to his dog and said, "Coffee Break, do your stuff." Coffee Break jumped to his feet, ate the cookies, drank the milk, dumped on the paper, sexually assaulted the other three dogs, claimed he injured his back while doing so, filed a grievance report for unsafe working conditions, put in for Workers Compensation and went home for the rest of the day on sick leave.

Things You'll Never Hear One Guy Say To Another Guy

> Does my butt look fat in this?
> I can't stop fantasizing about Dr. Ruth!
> I think those big, jacked-up trucks look ridiculous.
> I'm deeply offended by young women who go braless.
> I'm tired of beer. What say you to a nice, fruity Chablis?
> Our team lost 10-1. But we tried our best, and after all that's the important thing.
> There's nothing I like more than a quiet evening at home, watching a movie on Lifetime about some woman who gives up her baby and then suffers miserably.
> Want all my tools? I just realized I never do anything useful with them!
> You know what always makes me cry? Those long-distance commercials.
> Yours is bigger than mine.

The Cork

Two guys are in a locker room when one guy notices the other guy has a cork in his ass.

He says, "How'd you get a cork in your ass?"

The other guy says, "I was walking along the beach and I tripped over a lamp. There was a puff of smoke, and then a red man in a turban came oozing out. He said, "I am Tonto, Indian Genie. I can grant-um you one wish."

And I said, "No shit."

Request

A handsome hunk is jogging down the beach when he sees a girl in a wheelchair sitting on a pier crying. He runs over and asks why she's crying.

"I've never been kissed," she sobs.

So the hunk lifts her up, cradles her in his arms, and gives her a long, passionate kiss. "Now," he says, "you've been kissed." He puts her back in her chair and continues to run.

A week later, he's out jogging again when he sees the same girl on the same pier, crying again. "What is it this time?" he asks.

"I've never been screwed," the girl sobs.

Again, the hunk picks her up and cradles her gently. He slowly moves to the end of the pier, kissing her as he did the first time. Suddenly, he throws her as far out in the water as he can. "Now," he calls to her, "you're screwed."

Things That Suck About Being A Guy

- Even if you get you head caught in an industrial wood chipper, you're not allowed to cry.
- External genitalia are vulnerable to knees and fastballs.
- James Bond movies only come out every 2 years.
- No sofas in your restrooms.
- Ribbed for her pleasure - not yours.
- That Ferrari lists for over $300,000.
- You can't flirt your way out of a jam.
- You have to take out the garbage.
- You have to wear ties.
- "Women and children first."

Lumber Jack

An elderly man saw an ad in a magazine which said, "Lumberjacks needed in the Canadian North Woods." He flew up north and reported for the job.

The six foot three inch foreman looked down at the five foot two inch man and explained that they cut trees with axes and it takes brawn, which obviously he didn't have.

The old man said, "Just give me a test." So the foreman handed him an axe and said, "Go chop down that tree." The old man cut the tree down in four strokes, zip, zap, zoop, shlip. The foreman complimented him, but explained that he was being kind, but that wasn't the kind of trees they cut down, it was the kind they plant.

Then he said, pointing to a monster tree, those are the kind we cut down. The old man went over to the big tree and with ten strokes, zip, zap, zoop, shlipp, flop, blam, boom, crack, crunch, bim, he cut down the tree.

With that the foreman said, "You're hired, but where did you get the experience?"

The old man said, "In the Sahara Forest."

The foreman said, "You mean the Sahara Desert, don't you?"

And the old man said, "Sure, Now!"

Truck Driver

The state trooper was driving down the highway when, much to his surprise, he saw a truck driver pull over, walk to the side of his truck with a crowbar, bang on the side of truck several times, then continue on down the highway. Two miles down the road the trucker repeated the procedure, then did it again two miles farther on down.

Though the driver had not broken any laws, the patrolman's curiosity got the best of him; he pulled the man over and asked him to explain.

"It's simple," said the driver. "My load limit is two tons, and there are four tons worth of parakeets back there. If I don't keep half of them airborne, I'm in trouble."

Drinker

A well dressed gentlemen enters the bar of a five star restaurant, sits at the bar and orders four very expensive drinks. The bartender serves them on a silver tray, setting all four in front of the patron.

The man then consumes all four drinks in a matter of seconds. The bartender comments, "Wow, you sure must have a problem."

"If you had what I had," the man replies, "you'd drink them fast, too."

Leaning over, the sympathetic bartender asks, "What do you have?"

"Fifty cents," the man answers.

Q and A # 1

Q. What is the thinnest book in the world?
A. "What Men Know About Women."

Q. How many men does it take to screw in a light bulb?
A. One.. Men will screw anything.

Q. How does a man take a bubble bath?
A. He eats beans for dinner.

Q. Why do women rub their eyes when they wake up?
A. Because they don't have balls to scratch.

Q What's a man's idea of foreplay?
A. A half hour of begging.

Q. How can you tell if a man is sexually excited?
A. He's breathing

Q. What's the difference between men and government bonds?
A. Bonds mature.

Q. How do you save a man from drowning?
A. Take your foot off his head.

Q. What do men and beer bottles have in common?
A. They are both empty from the neck up.

Q. How can you tell if a man is happy?
A. Who cares?

Q. How many men does it take to change a roll of toilet paper?
A. We don't know. It never happens.

Q. How are men and parking spots alike?
A. The good ones are always taken, and the rest are handicapped.

Q. What's a man's idea of helping with the housework?
A. Lifting his leg so you can vacuum.

Q. What's the difference between men and E.T.?
A. E.T. phones home

Q. What does a man consider a seven course meal?
A. A hot dog and a six pack.

How Many Drinks?

Hank is having a drink in his local bar when in walks this gorgeous woman. Hank, not being too shy, goes up and sits next to her. He buys her a drink and then another and then another.

After this and the accompanying small-talk, Hank asks her back to his place for a "good time."

"Look," says the woman, "what do you think I am? I don't turn into a slut after 3 drinks, you know!"

"OK," replies Hank, "so how many does it take?"

Why Beer Is Better Than Women

1. A Beer always goes down easy.
2. A Beer doesn't care when you come home.

3. A Beer doesn't get jealous when you grab another Beer.
4. A Beer won't get upset if you come home and have another.
5. A frigid Beer is a good Beer.
6. After you have a Beer, the bottle is still worth 5 cents.
7. Beer doesn't demand equality.
8. Beer is always wet.
9. Beer is never late.
10. Beer labels come off without a fight.
11. Beer never has a headache.
12. Beer stains wash out.
13. Hangovers go away.
14. If you change Beers, you don't have to pay alimony.
15. If you pour a Beer right, you'll always get good head.
16. When you go to a bar, you know you can always pick up a Beer.
17. When your Beer goes flat, you toss it out.
18. You always know you're the first one to pop a Beer.
19. You can enjoy a Beer all month long.
20. You can have a Beer in public.
21. You can have more than one Beer in a night and not feel guilty.
22. You can share a Beer with your friends.
23. You don't have to wash a Beer before it tastes good.
24. You don't have to wine and dine Beer.
25. Your Beer will always wait patiently for you in the car while you play baseball.

Q and A # 2

Q. How can you tell soap operas are fictional?
A. In real life, men aren't affectionate out of bed.

Q. What should you give a man who has everything?
A. A woman to show him how to work it.

Q. Why do black widow spiders kill their males after mating?
A. To stop the snoring before it starts.

Q. Why don't men have mid-life crises?
A. They stay stuck in adolescence.

Q. How does a man show he is planning for the future?
A. He buys two cases of beer instead of one.

Q. How is being at a singles bar different from going to the circus?
A. At the circus the clowns don't talk.

Q. What makes men chase women they have no intention of marrying?

A. The same urge that makes dogs chase cars they have no intention of driving.

Q. What do you do with a bachelor that thinks he is God's gift?
A. Exchange him.

Q. Why do bachelors like smart women?
A. Opposites attract.

Q. Why are husbands like lawn mowers?
A. They are hard to get started, emit foul odors, and don't work half the time.

Q. What's the difference between a new husband and a new dog?
A. After a year, the dog is still excited to see you.

Q. What's a man's idea of foreplay?
A. A half hour of begging.

Q. What's the difference between men and government bonds?
A. Bonds mature.

Q. What do men and beer bottles have in common?
A. They're both empty from the neck up.

Q. How many men does it take to change a roll of toilet paper?
A. We don't know-it's never happened.

Q: Why are men like laxatives?
A: They irritate the shit out of you

Q: Why did God create man?
A: Because vibrators don't mow lawns

Q: What are two reasons men don't mind their own business?
A: No mind-No business

Q: How are men and parking spots alike?
A: The good ones are taken and what's left is handicapped

Q: Why is it hard for a women to find men who are sensitive, caring and good looking?
A: Because those men already have boyfriends

Q: How is a man like a snowstorm?

A: You never know when he's coming, how many inches you'll get or how long it will last

Q: Why are men given larger brains than dogs?
A: So they don't hump women's legs at cocktail parties

Q: Why can't men get mad cow disease?
A: Because they're all pigs

Men Are Like...

> Men are like cement... after getting laid, they take a long time to get hard.
> Men are like chocolate bars... sweet, smooth, and they usually head right for your hips.
> Men are like coffee... the best ones are rich, warm, and can keep you up all night long.
> Men are like computers...hard to figure out and never have enough memory.
> Men are like coolers.... load them with beer and you can take them anywhere
> Men are like department stores... their clothes should always be half off.
> Men are like horoscopes... they always tell you what to do and are usually wrong.
> Men are like plungers... they spend most of their lives in a hardware store or the bathroom.
> Men are like vacations... they never seem to be long enough.

Midgets

A guy is standing at a urinal when he notices that he's being watched by a midget. Although the little fellow is staring at him intently, the guy doesn't get uncomfortable until the midget drags a small stepladder up next to him, climbs it, and proceeds to admire his privates at close range.

"Wow," comments the midget, "Those are the nicest balls I have ever seen!"

Surprised-and flattered-the man thanks the midget and starts to move away.

"Listen, I know this is a rather strange request," says the little fellow, "but I wonder if you would mind if I touched them."

Again the man is rather startled, but seeing no real harm in it, he obliges the request. The midget reaches out, gets a tight grip on the man's balls, and says, "Okay, hand me your wallet or I'll jump off the ladder!"

Manspeak Translations

"But I hate to go shopping."
Really means... "Because I always wind up outside the dressing room holding your purse."

"No, I left plenty of gas in the car."
Really means... "You may actually get it to start."

"I'm going to stop off for a quick one with the guys."
Really means... "I am planning on drinking myself into a vegetative stupor with my chest pounding, mouth breathing, pre-evolutionary companions."

"I heard you."
Really means... "I haven't the foggiest clue what you just said, and am hoping desperately that I can fake it well enough so that you don't spend the next 3 days yelling at me."

"You know I could never love anyone else."
Really means..."I am used to the way you yell at me, and realize it could be worse."

"You look terrific."
Really means..."Oh, God, please don't try on one more outfit. I'm starving."

"I brought you a present."
Really means..."It was free ice scraper night at the ball game."

"I missed you."
Really means..."I can't find my sock drawer, the kids are hungry and we are out of toilet paper."

"I'm not lost. I know exactly where we are."
Really means... "No one will ever see us alive again."

"We share the housework."
Really means... "I make the messes, she cleans them up."

"This relationship is getting too serious."
Really means..."I like you more than my truck."

"I recycle."
Really means..."We could pay the rent with the money from my empties."

"Of course I like it, honey, you look beautiful."
Really means... "Oh, man, what have you done to yourself?"

"It sure snowed last night."
Really means..."I suppose you're going to nag me about shoveling the walk now."

"It's good beer."
Really means..."It was on sale."

"I don't need to read the instructions."
Really means..."I am perfectly capable of screwing it up without printed help."

"I'll fix the garbage disposal later."
Really means..."If I wait long enough you'll get frustrated and buy a new one."

"I'll take you to a fancy restaurant."
Really means..."Someplace that doesn't have a drive-thru window."

"I broke up with her."
Really means... "She dumped me."

College Costs

On the first day of college, the dean/principal addresses the students pointing out some of the rules. "The female dormitory will be out-of-bounds for all male students, so too the male dormitory to the female students.

Anybody caught breaking this rule will be fined $20 the first time. Anybody caught breaking this rule the second time will be fined $60. Being caught a third time will incur a hefty fine of $180. Are there any questions?."
To this, a male student in the crowd inquires, "How much for a season pass?"

Missing Car

A man walks out of a bar, stumbling back and forth with a key in his hand. A cop on the beat sees him and approaches, Can I help you, sir?
"Yesssh! Sshomebody sshtole my car!" the man replies.
The cop asks, "Where was your car the last time you saw it?"
"It wasssh at the end of thisssh key!" replied the man.

About this time the cop looks down to see that the man's member is being exhibited for all the world to see. He asks the man, "Sir, are you aware that you are exposing yourself?"

The man looks down woefully and without missing a beat, moans "Ohhh GOD...they got my girlfriend too!"

Wink

A man with a winking problem is applying for a position as a sales representative for a large firm. The interviewer looks over his papers and says, "This is phenomenal. You've graduated from the best schools; your recommendations are wonderful, and your experience is unparalleled. Normally, we'd hire you without a second thought. However, a sales representative has a highly visible position, and we're afraid that your constant winking will scare off potential customers. I'm sorry, we can't hire you."

"But wait," he said. "If I take two aspirin, I'll stop winking!"

"Really? Great! Show me!" said the interviewer.

So the applicant reaches into his jacket pocket and begins pulling out all sorts of condoms: red condoms, blue condoms, ribbed condoms, flavored condoms; finally, at the bottom, he finds a packet of aspirin. He tears it open, swallows the pills, and stops winking.

"Well," said the interviewer, "that's all well and good, but this is a respectable company, and we will not have our employees womanizing all over the country!"

"Womanizing? What do you mean?" Exclaimed the man. "I'm a happily married man!"

"Well then, how do you explain all these condoms?" the interviewer says while pointing to the pile on the desk.

"Oh, that," he sighed. "Have you ever walked into a pharmacy, winking, and asked for aspirin?"

Traveling Salesman

A traveling salesman is in a small town in the Midwest, when his trip is suddenly prolonged for an extra month. He was already getting bored there and over the course of the extra month he becomes very homesick.

Finally, he decides to give in to temptation and visit the local brothel. He walks up to the madam and hands her a hundred dollars and says, "Give me the worst blowjob in town."

The madam says, "For this kind of money, you can have the best blowjob."

"No, no," says the man, "You don't understand, I'm not horny, I'm homesick."

Lucky Frog

A man takes the day off work and decides to go out golfing. He is on the second hole when he notices a frog sitting next to the green. He thinks nothing of it and is about to shoot when he hears, "Ribbit. 9 Iron" The man looks around and doesn't see anyone. "Ribbit. 9 Iron." He looks at the frog and decides to prove the frog wrong, puts his other club away, and grabs a 9 iron. Boom! he hits it 10 inches from the cup. He is shocked. He says to the frog, "Wow that's amazing. You must be a lucky frog, eh?" The frog reply's "Ribbit. Lucky frog."

The man decides to take the frog with him to the next hole. "What do you think frog?" the man asks. "Ribbit. 3 wood." The guy takes out a 3 wood and Boom! Hole in one. The man is befuddled and doesn't know what to say. By the end of the day, the man golfed the best game of golf in his life and asks the frog, "OK where to next?" The frog reply, "Ribbit. Las Vegas."

They go to Las Vegas and the guy says, "OK frog, now what?" The frog says, "Ribbit Roulette." Upon approaching the roulette table, the man asks," What do you think I should bet?" The frog replies, "Ribbit. $3000,black 6." Now, this is a million-to-one shot to win, but after the golf game, the man figures what the heck. Boom! Tons of cash comes sliding back across the table. The man takes his winnings and buys the best room in the hotel.

He sits the frog down and says, "Frog, I don't know how to repay you. You've won me all this money and I am forever grateful." The frog replies, "Ribbit, Kiss Me." He figures why not, since after all the frog did for him he deserves it. With a kiss, the frog turns into a gorgeous 15-year-old girl.

"And that, your honor, is how the girl ended up in my room."

From another Angle

> A woman's work that is never done is the stuff she asked her husband to do.
> Go for younger men. You might as well — they never mature anyway.
> Husbands are like children — they're fine if they're someone else's.
> If you think the way to a man's heart is through his stomach you're aiming too high.
> If you want a nice man go for a bald one — they try harder.
> Men are all the same — they just have different faces so you can tell them apart.
> Men are like animals — messy, insensitive and potentially violent - but they make great pets.
> Men's brains are like the prison system — not enough cells per man.

➢ Never trust a man who says he's the boss at home. He probably lies about other things too.

➢ Scientists have just discovered something that can do the work of five men — a woman.

➢ The best reason to divorce a man is a health reason: you're sick of him.

➢ There are a lot of words you can use to describe men — strong, caring, loving they'd be wrong but you could still use them.

➢ There are only two four letter words that are offensive to men - "don't" and "stop."

➢ Whenever you meet a man who would make a good husband, you will usually find that he is.

➢ Woman don't make fools of men — most of them are the do-it-yourself types.

Women

Tight Skirt

A woman is trying to board a bus, but her skirt is too tight and she can't step up. She reaches behind her and lowers the zipper a bit and tries again. Her skirt's still too tight. She reaches behind her and lowers the zipper some more, but she still can't get on, and lowers the zipper a third time.

Suddenly, she feels two hands on her butt, trying to push her up onto the bus. She spins around and says, "Sir, I don't know you well enough for you to do that!"

To which he replies, "Lady, I don't know you well enough for you to unzip my fly three times so I guess we are even."

Differences Between Good Girls And Bad Girls

➢ Good girls loosen a few buttons when its hot	➢ Bad girls make it hot by loosening a few buttons
➢ Good girls wax their floors	➢ Bad girls wax their bikini line
➢ Good girls blush during sex scenes in movies	➢ Bad girls know they could do it better
➢ Good girls wear white cotton panties	➢ Bad girls don't wear any
➢ Good girls pack their toothbrush	➢ Bad girls pack their diaphragms
➢ Good girls own only one credit card	➢ Bad girls own only one bra and rarely use it
➢ Good girls wear high heels to work	➢ Bad girls wear high heels to bed
➢ Good girls prefer the missionary position	➢ Bad girls do to, but only for starters
➢ Good girls say no	➢ Bad girls say when?

Sex Drive

Scientists have discovered a food that diminishes a woman's sex drive by 90 percent. ... wedding cake!

Port vs. Sherry

A wealthy playboy met a beautiful young girl in an exclusive lounge. He took her to his lavish apartment where he soon discovered she was not a tramp, but was well groomed and apparently very intelligent. Hoping to get her into bed he began showing her his collection of expensive paintings, first editions by famous authors and offered her a glass of wine.

He asked whether she preferred Port or Sherry and she said, "Oh, Sherry by all means. To me it's the nectar of the gods. Just looking at it in a crystal-clear decanter fills me with a glorious sense of anticipation. When the stopper is removed and the gorgeous liquid is poured into my glass, I inhale the enchanting aroma and I'm lifted on the wings of ecstasy. It seems as though I'm about to drink a magic potion and my whole being begins to glow. The sound of a thousand violins being softly played fills my ears and I'm transported into another world.

"Jackpot!" thought the playboy.

"On the other hand," she added, "Port makes me fart."

Blind

Then there was the nicely endowed woman who was in her office on a hot summer day.

The heat really was unbearable. Since she was in her office, and no one was there, she took off her clothes exposing her breasts.

A few minutes later, her secretary buzzed her on the phone and said, "the blind man is here." She thought for a second and decided to allow the blind man to come in. She figured since he was blind she would not need to get dressed.

When he entered, he said, "WOW! Those are the most beautiful breasts I've ever seen. Now, where do you want me to put the blinds?"

Reasons Why Cucumbers Are Better Than Men

1. A cucumber doesn't care if you're a virgin.
2. A cucumber will always respect you in the morning.
3. A cucumber won't ask: "Am I the first?."
4. A cucumber won't drag you to a John Wayne Film Festival.
5. A cucumber won't eat all the popcorn... or send you out for Milk Duds.
6. Cucumbers are easy to pick up.
7. Cucumbers can get away any weekend.

8. Cucumbers can handle rejection.
9. Cucumbers don't get too excited.
10. Cucumbers never suffer from performance anxiety.
11. Cucumbers never want to get it on when your nails are wet.
12. Cucumbers stay hard for a week.
13. Cucumbers won't care what time of the month it is.
14. Cucumbers won't pout if you have a headache.
15. Cucumbers won't tell anyone you're not a virgin anymore.
16. Cucumbers won't tell other cucumbers you're a virgin.
17. Cucumbers won't tell you size doesn't count.
18. The average cucumber is at least 6 inches long.
19. With a cucumber you can always wait until you get home.
20. With a cucumber you can get a single room... and you won't have to check-in as Mrs. Cucumber.
21. With a cucumber you don't have to be a virgin more than once.
22. With a cucumber you never have to say you're sorry.
23. You can fondle a cucumber in a supermarket... and you know how firm it is before you take it home.
24. You can go to a drive-in with a cucumber... and you can stay in the front seat.
25. You can go to a movie with a cucumber... and see the movie.

Horse Ride

A lady gets stranded in the desert. She is waiting for about an hour until an Indian rides by on his horse. She asked him for a ride into town.

"Hop on" he says. She jumps on the back of his horse and they take off.

About 2 minutes after riding the Indian lets out a "YAHOO." Then a few minutes later he yells it again. This happens the whole way into town. When they reach town, the Indian lets out a really loud yahoo.

As the lady is getting off the horse a store owner asked her what she did to the Indian. "Nothing", she replied. I just sat on the back of the horse and held on to the saddle horn."

"That explains it, "the store owner replied. "Indians don't use saddles."

Facts about Women # 1

1. Women love to shop. It is the one area of the world where they feel like they're actually in control.
2. Women especially love a bargain. The question of "need" is irrelevant, so don't bother pointing it out. Anything on sale is fair game.
3. Women never have anything to wear. Don't question the racks of clothes in the closet; you "just don't understand."
4. Women need to cry. And they won't do it alone unless they know you can hear them.

5. Women will always ask questions that have no right answer, in an effort to trap you into feeling guilty.
6. Women love to talk. Silence intimidates them and they feel a need to fill it, even if they have nothing to say.
7. Women need to feel like there are people worse off than they are. That's why soap operas and Oprah Winfrey-type shows are so successful.
8. Women don't need sex as often as men do. This is because sex is more physical for men and more emotional for women. Just knowing that the man "wants" to have sex with them fulfills the emotional need.
9. Women hate bugs. Even the strong-willed ones need a man around when there's a spider or a wasp involved.
10. Women cannot keep secrets. They eat away at them from the inside. And they don't view it as being untrustworthy, providing they only tell two or three people.
11. Women always go to public restrooms in groups. It gives them a chance to gossip.
12. Women cannot refuse to answer a ringing phone, no matter what she's doing. It might be the lottery calling.
13. Women never understand why men love toys. Men understand that they wouldn't need toys if women had an "on/off" switch.
14. Women think all beer is the same.
15. Women keep three different shampoos and two different conditioners in the shower. After a woman showers, the bathroom will smell like a tropical rain forest.
16. Women don't understand the appeal of sports. Men seek entertainment that allows them to escape reality. Women seek entertainment that reminds them of how horrible things "could" be.
17. If a man goes on a seven-day trip, he'll pack five days worth of clothes and will wear some things twice; if a woman goes on a seven-day trip she'll pack 21 outfits because she doesn't know what she'll feel like wearing each day.
18. Women brush their hair "before" bed.
19. Watch a woman eat an ice cream cone and you'll have a pretty good idea about how she'll be in bed.
20. Women are paid less than men, except for two fields: Modeling and Divorce.
21. Women are "never" wrong. Apologizing is the man's responsibility, "It's there in the Bible." Hmmm, who was it that gave Adam the apple?
22. Women do *not* know anything about cars. "Oil-stick, oil doesn't stick?"
23. Women have better restrooms. They get the nice chairs and red carpet. Men just get a large bowl to share.

24. The average number of items in a typical woman's bathroom is 437. A man would not be able to identify most of these items.
25. Women love cats. Men say they love cats, but when women aren't looking, men kick cats.

Kuwait

A journalist had done a story on gender roles in Kuwait several years before the Gulf War, and she noted then that women customarily walked about 10 feet behind their husbands.

She returned to Kuwait recently and observed that the men now walked several yards behind their wives. She approached one of the women for an explanation.

"This is marvelous," said the journalist. "What enabled women here to achieve this reversal of roles?"

Replied the Kuwaiti woman: "Land mines."

Things you'll never hear one woman say to another woman

> He earned more than I do, so I broke up with him.
> He talks our relationship to death! It's making me crazy!
> His new girlfriend is thinner and better-looking than I am, and I'm happy for them both.
> I just realized-my butt doesn't look fat in this-my butt is fat!
> If he doesn't let me hold the remote, I get all moody.
> I'm sick of dating doctors and lawyers! Give me a good old-fashioned waiter with a heart of gold any day!
> Oh, look, that women and I have the same dress on! I think I'll go introduce myself!
> That swimsuit really flatters your figure! Would you mind keeping my husband company while I go for a swim?
> We're redecorating the bedroom, and he keeps bugging me to help him with the color choices!
> Why can't I find a guy who'll have a wild carefree night of sex and then just go his separate way for once?

Vaseline Use

A woman answers the door to a market researcher. "Good morning madam, I'm doing some research for Vaseline. Do you use it at all in your household?"

"Oh yes, all the time. It's very good for cuts, grazes and burns."

"Do you use it for anything else?"

"Like what?"

"Ahem.. err.. well.. during.. ahem.. sex."

"Oh, of course. Yes, we smear it on the bedroom doorknob to keep the kids out."

Tests Show Beer Contains Female Hormones

Yesterday scientists in the USA revealed that beer contains small traces of female hormones.

To prove their theory they fed 100 men 12 pints of beer and observed that 100% of them lost all sense of reasoning, started talking nonsense, and couldn't drive.

Horny Widows

Down in Florida, two widows were talking and one asked the other, "Do you ever get to feeling horny?" "Yes," her friend replied. "What do you do about it?"

"I usually suck on a Lifesaver." After a moment of stunned silence her friend asked, "Well, what beach do you go to?"

Married Ten Times

I have been married ten times. Let me tell you what is wrong with each man:

> ➤ My 1st husband was a musician. All he wanted to do was play with it.
> ➤ My 2nd husband was a doctor. All he wanted to do was examine it.
> ➤ My 3rd husband was a politician. All he wanted to do was make promises to it.
> ➤ My 4th husband was a salesman. All he wanted to do was to talk to it.
> ➤ My 5th husband was a policeman. All he wanted to do was keep it under lock and key.
> ➤ My 6th husband was a photographer. All he wanted to do was to take pictures of it.
> ➤ My 7th husband was a hairdresser. All he wanted to do was to tease it.
> ➤ My 8th husband was a gourmet chef. All he wanted to do was taste it.
> ➤ My 9th husband was a preacher. All he wanted to do was save it.
> ➤ My 10th husband and last is the man I am still married to and by far the best. He is a mechanic. He tore it up the first night and he has been working on it ever since.

Nicknames

Three women were sitting around one night talking about their boyfriends when they decided they would give their men nicknames based on types of soda.

The first woman said: "I'm gonna call Tom "Mountain Dew" because he is as strong as a mountain and always wants to do it!"

The second woman said: "I'm gonna call Bruce "7-Up" because he has seven inches and it is always up!"

The third woman said: "I'm gonna call my man "Jack Daniels.""

The other two women responded: "Jack Daniels? But that's a hard liquor."

The third woman replied: "That's my husband!"

What Woman Say About Men

- ➢ A man's idea of serious commitment is usually, "oh alright, I'll stay the night."
- ➢ Definition of a bachelor; a man who has missed the opportunity to make some woman miserable.
- ➢ Don't imagine you can change a man - unless he's in nappies.
- ➢ Go for younger men. You might as well - they never mature anyway.
- ➢ If he asks what sort of books you're interested in, tell him check books.
- ➢ If he asks you if you're faking it tell him no, you're just practicing.
- ➢ If they put a man on the moon - they should be able to put them all there.
- ➢ If you want a committed man look in a mental hospital.
- ➢ Love is blind, but marriage is a real eye-opener.
- ➢ Men are all the same - they just have different faces so you can tell them apart.
- ➢ Never do housework. No man ever made love to a woman because the house was spotless.
- ➢ Never let your man's mind wander - it's too little to be let out alone.
- ➢ Never sleep with a man who's named his willy.
- ➢ Remember a sense of humor does not mean that you tell him jokes, it means you laugh at his.
- ➢ Remember you are known by the idiot you accompany.
- ➢ Sadly, all men are created equal.
- ➢ So many men - so many reasons not to sleep with any of them.
- ➢ Tell him you're not his type - you have a pulse.
- ➢ The best way to get a man to do something is to suggest they are too old for it.

> ➤ The children of Israel wandered around the desert for 40 years. Even in biblical times men wouldn't ask for directions.
> ➤ The main point of having a boyfriend is so that he can one day graduate to the exalted status of a "former boyfriend."
> ➤ The only reason men are on this planet is that vibrators cannot dance or buy drinks.
> ➤ There are a lot of words that you can use to describe - strong, caring, loving - they'd be wrong - but you could still use them.
> ➤ There are two significant influences in a man's life and they are both his mother.
> ➤ What do you do if your boyfriend walks out? You shut the door.
> ➤ When he asks you if he's your first tell him, "You may be, you look familiar."
> ➤ Women don't make fools of men - most of them are the do-it-yourself types.
> ➤ Women sleep with men, who if they were women, they wouldn't even have bothered to have lunch with.

Closure

This lady died. She had been quite the gregarious and amorous sort for many years...you get the picture. Whole lot a love. Whole lot a husbands, too.

All her girlfriends came to the funeral. As they lowered her coffin into the grave, one of the girls sighed, "Finally they are together."

"What?", asked another lady, "which one of her husbands is she joining here?"

"Oh, I'm not talking about her husbands," explained the friend. "I'm talking about her legs."

Facts about Women # 2

1. "Oh, nothing," has an entirely different meaning in woman-language than it does in man-language.
2. A woman will dress up to go shopping, water the plants, empty the garbage, answer the phone, read a book, or get the mail.
3. All women are overweight by definition; don't agree with them about it. Women always have 5 pounds to lose, but don't bring this up unless they really have 5 pounds to gain.
4. If a man ticks off a woman she will often respond by getting a fuzzy toilet cover which warms their rear, but makes it impossible for the lid to stay up thus it constantly gets peed on by the guys. (which gets them in more trouble)
5. If it is not Valentines day and you see a man in a flower shop, you can probably start up a conversation by asking, "What did you do?"

6. It's okay for women to dance with each other and not be gay, You don't see straight men dancing together.
7. Lewis Carroll's Caterpillar had nothing on women.
8. Only women understand the reason for "guest towels" and the "good china."
9. PMS stands for: Permissible Man-Slaughter. (Or at least men think it means that. PMS also stands for Preposterous Mood Swings and Punish My Spouse.
10. The first naked man women see is "Ken."
11. The most embarrassing thing for women is to find another woman wearing the same dress at a formal party. You don't hear men say, "Oh-my-GOD, there's another man wearing a black tux, get me out of here!"
12. Women are insecure about their weight, butt, and breast sizes.
13. Women can get out of speeding tickets by pouting. This will get men arrested.
14. Women cannot use a map without turning the map to correspond to the direction that they are heading.
15. Women do NOT want an honest answer to the question, 'How do I look?'
16. Women don't really care about a sense of humor in a guy despite claims to the contrary. You don't see women trampling over Tom Cruise to get to Gilbert Gottfried, do you?
17. Women don't try as hard as men during sex; after all, they don't fall asleep afterwards.
18. Women fake orgasm because men fake foreplay.
19. Women love to talk on the phone. A woman can visit her girlfriend for two weeks, and upon returning home, she will call the same friend and they will talk for three hours.
20. Women never check to see if the lid is up. They seem to prefer taking a flying butt leap towards the bowl and then chewing men out because they "left the seat up" instead of taking two seconds and lowering it themselves.
21. Women want equal rights, but you rarely hear them clamoring to be let into the draft to cover the responsibilities that go with those rights. All women seek equality with men until it comes to sharing the closet, taking out the trash, and picking up the check.
22. Women will drive miles out of their way to avoid the possibility of getting lost using a shortcut.
23. Women will make three right-hand turns to avoid making one left-hand turn.
24. Women will spend hours dressing up to go out, and then they'll go out and spend more time checking out other women. Men can never catch women checking out other men; women will always catch men checking out other women.

Kids

Nightwalker

A little boy gets up to go to the bathroom in the middle of the night. As he passes his parent's bedroom he peeks in through the keyhole. He watches for a moment, then continues on down the hallway, saying to himself, "Boy, and she gets mad at me for sucking my thumb"

Night Sounds

This little boy wakes up 3 nights in a row when he hears a thumping sound coming from his parents bedroom. Finally one morning he goes to his mom and says, "Mommy, every night I hear you and daddy making noise and when I look in you're bouncing up and down on him."

His mom is taken by surprise and says." Oh... well I'm bouncing on his stomach because he's fat and that makes him thin again."

And the boy says, "That won't work." His mom says, "Why?." And the boy replies. "Because the lady next door comes by after you leave each day and blows him back up!"

Dance Date

It's the Spring of 1957 and Bobby goes to pick up his date. He's a pretty hip guy with his own car. When he goes to the front door, the girl's father answers and invites him in. "Carrie's not ready yet, so why don't you have a seat?," he says.

"That's cool" says Bobby.

Carrie's father asks Bobby what they're planning to do.

Bobby replies politely that they will probably just go to the soda shop or a movie.

Carrie's father responds "why don't you two go out and screw? I hear all the kids are doing it."

Naturally, this comes as a quite a surprise to Bobby - so he asks Carrie's Dad to repeat it.

"Yeah," says Carrie's father, "Carrie really likes to screw; she'll screw all night if we let her!"

Well, this just made Bobby's eyes light up, and his plan for the evening was beginning to look pretty good.

A few minutes later, Carrie comes downstairs in her little poodle skirt and announces that she's ready to go. Almost breathless with anticipation, Bobby escorts his date out the front door.

About 20 minutes later, Carrie rushes back into the house, slams the door behind her, and screams at her father: "DAMMIT DADDY! IT'S CALLED THE TWIST!!!"

118

Blackboard

One day when the teacher walked into the classroom, she noticed that someone had written the word 'PENIS' (in tiny letters) on the blackboard. She scanned the class looking for a guilty face. Finding none, she rubbed the word off and began class.

The next day, the word 'PENIS' was written on the board again; this time it was written about halfway across the board. Again she looked around in vain for the culprit, so she proceeded with the day's lesson.

Every morning for about a week, she went into the classroom and found the same disgusting word written on the board, each day's being larger than the previous one, and each being rubbed off vigorously.

At the end of the second week, she walked in expecting to be greeted by the same word on the board but instead found the words: "The more you rub it, the bigger it gets."

Taxi Ride

A mother, accompanied by her small daughter, were in New York City. The mother was trying to hail a cab, when her daughter noticed several wildly dressed women who were loitering on a nearby street corner.

The mother finally hailed her cab and they both climbed in, at which point the daughter asks her mother, "Mummy, what are all those ladies waiting for by that corner?", to which the mother replies, "Those ladies are waiting for their husbands to come home from work."

The cabbie, upon hearing this exchange, turns to the mother and says, "Ahhhhhhh, C'mon lady!!!! Tell your daughter the truth!!!! For crying out loud. They're hookers!"

A brief period of silence follows, and the daughter then asks, "Mummy, do the ladies have any children?"

The mother replies, "Of course dear. Where do you think cabbies come from?"

Thumbs

Husband and wife and their two sons are watching TV. She looks at her husband and winks at him, he gets the message and says, "Excuse us for a few minutes boys, we're going up to our room for a little while."

Pretty soon one of the boys becomes curious, goes upstairs and sees the door to his parents bedroom is ajar. He peeks in for a few minutes, trots downstairs, gets his little brother and takes him up to peek into the bedroom.

"Before you look in there," he says, "keep in mind this is the same woman who paddled our butts for sucking our thumbs."

My Dad, Your Dad # 1

Two kids were having the standard argument about whose father could beat up whose father.

One boy said, "My father is better than your father."

The other kid said, "Well, my mother is better than your mother."

The first boy paused, "I guess you're right. My father says the same thing."

Questions

A boy goes up to his mother and say, "Mommy, how old are you?" The mom says, "Son, there are some questions you should never ask a lady, and that's one of them."

The boy says. "Okay, Mommy. How much do you weigh?" She says, "Son, that's another question you should never ask a lady."

The boy says. "Okay, Mommy. Why did you and daddy get a divorce?" The mom says, "Okay, that's enough questions for now. Why don't you go outside and play?"

The boy goes outside and sees a friend of his. He says to the friend, "I kept asking my mom all of these questions about herself, and she wouldn't answer them." The friend says, "You know what you should do? You should go into her purse and look at her driver's license. They have all sorts of neat information on those."

The boy thinks it's a good idea. He goes into the house and gets his mom's driver's license, then looks it over. He goes to his mom, and says, "Guess what, mom? I looked at your driver's license, and now I know how old you are!"

The mom says, "Really?" The boy says, "Yeah, you're 32! And I know how much you weigh, too!"

The mom says, "Really?" The boy says, "Yeah, you weigh 150 pounds. And I also know why daddy divorced you, too!"

The mom says, "You do?" The boy says, "Yeah, because you got an 'F' in sex!"

Little Johnny

Little Johnny greeted his mother at the door after she had been out of town all week and said, "Mommy, guess what? Yesterday, I was playing in the closet in your bedroom and Daddy came into the room with the lady from next door. They undressed and got into bed and then Daddy got on top of her and..."

The mother held up her hand and said, "Not another word! Wait until your father gets home and then I want you to tell him exactly what you've just told me."

The father came home and the wife tells him that she's leaving him.

"But why?" croaked the husband.

"Go ahead, Johnny. Tell Daddy what you've just told me."

"Well," said little Johnny, "I was playing in your closet and Daddy came upstairs with the lady next door and they got undressed and they got into bed and Daddy got on top of her and they did just what you did, Mommy, with Uncle Bob."

Wishing

A few months after his parents were divorced, little Johnny passed by his mom's bedroom and saw her rubbing her body and moaning, "I need a man, I need a man!" Over the next couple of months, he saw her doing this several times.

One day, he came home from school and heard her moaning. When he peeked into her bedroom, he saw a man on top of her. Little Johnny ran into his room, took off his clothes, threw himself on his bed, started stroking himself, and moaning, "Ohh, I need a bike! I need a bike!"

Daughters

A worried father confronted his daughter one night. "I don't like that new boyfriend, he's rough and common and bloody stupid with it."

"Oh no, Daddy," the daughter replied, "Fred's ever so clever, we've only been going out nine weeks and he's cured me of that illness I used to get once a month."

Death

A teacher was asking her class what their fathers did. When she asked young Johnny, he said, "My father's dead, Miss."

"Oh, I am sorry, Johnny. In that case, what did he do before he died?"

"He went blue and collapsed."

My Dad, Your Dad # 2

Three boys are in the schoolyard bragging of how great their fathers are.

The first one says: "Well, my father runs the fastest. He can fire an arrow, and start to run, I tell you, he gets there before the arrow."

The second one says: "Ha! You think that's fast! My father is a hunter. He can shoot his gun and be there before the bullet."

The third one listens to the other two and shakes his head. He then says: "You two know nothing about fast. My father is a civil servant. He stops working at 4:30 and he is home by 3:45."

Fascinating

The teacher asked her students to use the word "fascinate" in a sentence. Mary said, "My family went to the New Your City Zoo, and we saw all the animals. It was fascinating." The teacher said, "That was good, but I wanted the word "fascinate."

Sally raised her hand. She said, "My family went to the Philadelphia Zoo and saw the animals.

I was Fascinated." The teacher said, "good, but I wanted the word "fascinate."

Little Billy raised his hand. The teacher hesitated because Billy was noted for his bad language. She finally decided there was no way he could damage the word "fascinate" so she called on him.

Billy said, "My sister has a sweater with 10 buttons, but her boobs are so big she can only "fasten 8."

Frog Princess

A boy was crossing a road one day when a frog called out to him and said, "If you kiss me, I'll turn into a beautiful princess."

He bent over, picked up the frog and put it in his pocket. The frog spoke up again and said, "If you kiss me and turn me back into a beautiful princess, I will stay with you for one week."

The boy took the frog out of his pocket, smiled at it and returned it to the pocket. The frog then cried out, "If you kiss me and turn me back into a princess, I'll stay with you and do ANYTHING you want."

Again the boy took the frog out, smiled at it and put it back into his pocket.

Finally, the frog asked, "What is the matter? I've told you I'm a beautiful princess, that I'll stay with you for a week and do anything you want. Why won't you kiss me?"

The boy said, "Look I'm a software engineer. I don't have time for a girlfriend, but a talking frog is cool."

Question

A little boy asked his father, "Daddy, how much does it cost to get married?"

And the father replied, "I don't know, son, I'm still paying for it."

Jesus

A Sunday School teacher of pre-schoolers was concerned that his students might be a little confused about Jesus Christ because of the Christmas season's emphasis on His birth. He wanted to make sure they understood that the birth of Jesus occurred a long time ago, that He grew up, etc.

So he asked his class, "Where is Jesus today?"

Steven raised his hand and said, "He's in heaven."

Mary was called on and answered, "He's in my heart."

Little Johnny, waving his hand furiously, blurted out, "I know! I know! He's in our bathroom!!!" The whole class got very quiet, looked at the teacher, and waited for a response.

The teacher was completely at a loss for a few very long seconds. He finally gathered his wits and asked Little Johnny how he knew this.

And Little Johnny said, "Well... every morning, my father gets up, bangs on the bathroom door, and yells 'Jesus Christ, are you still in there?'!"

Science Class

A science teacher set up a simple experiment to show her class the danger of alcohol. She set up 2 glasses, one containing water, the other containing gin. Into each she dropped a worm.

The worm in the water swam merrily around. The worm in the gin quickly died.

"What does this experiment prove?" she asked.

Little Johnny from the back row piped up: "It proves that if you drink gin you won't have worms"

Laws

Universal Laws

Asimov's First Law Of Observation
> When somebody you greatly admire and respect appears to be thinking deep thoughts, they probably are thinking about lunch.

Campbell's First Law Of Planning
> To estimate the time it takes to do a task, estimate the time you think it should take, multiply by 2, and change the unit of measure to the next highest unit, thus we allocate 2 days for a one-hour task.

Carter's Observation
> An elephant is a mouse built to government specifications.

Claire Booth Luce's Observation On Good Deeds
> No good deed will go unpunished.

Fenkell's Third Principle
> You can lead a horse to water, but if you can get him to float on his back, you've got something.

First Law Of Strolling
> Never step in anything soft.

Fraser's Rule
> Warranty and guarantee clauses are voided by payment of the invoice.

Goodrich's Postulate
> Anything based on greed and avarice is on a firm footing and will prevail.

Homer's Law
> You cannot fall off the floor.

Law Of International Travel
> Never, ever, fly on the airline of the country from which you are departing.

Law Of Mathematical Uncertainty
> If mathematically you end up with the incorrect answer, try multiplying by the page number.

Law Of Probability
Dispersal Whatever it is that hits the fan will not be evenly distributed.

Mitchell's First Law
You cannot successfully determine beforehand which side of the bread to butter.

Mitchell's Second Law
The chance of the bread falling with the buttered side down is directly proportional to the cost of the carpet.

Mrs Murphy's Law
If anything can go wrong, it will go wrong when he's out of town.

Pessimists' First Law
The light at the end of the tunnel is the headlamp of an oncoming train.

Sitting Bull's Observation
Give me a home where the buffalo roam, and you've got a room full of buffalo chips.

Toddler Property Laws

> If I like it, it's mine.
> If it's in my hand, it's mine.
> If I can take it from you, it's mine.
> If I had it a little while ago, it's mine.
> If it's mine, it must never appear to be yours in any way.
> If I'm doing or building something, all the pieces are mine.
> If it looks just like mine, it's mine.
> If I think it's mine, it's mine.
> If it's yours and I steal it, it's mine.
> If I ... Whoops! Sorry! I goofed! Instead of reading the Toddler Property Laws, I've been reading Microsoft's Business Plan.

Things that sound dirty in law

> Better leave the handcuffs on.
> Can you get him to drop his suit?
> Counselor, let's do it in chambers.
> For $200 an hour, she better be good!
> Have you looked through her briefs?
> He is one hard judge!
> His attorney withdrew at the last minute.
> Is it a penal offense?
> The judge gave her the stiffest one he could.

> ➤ Think you can get me off?

Laws Of Work

1. A pat on the back is only a few centimeters from a kick in the butt.
2. After any salary raise, you will have less money at the end of the month than you did before.
3. Anyone can do any amount of work provided it isn't the work he/she is supposed to be doing.
4. At work, the authority of a person is inversely proportional to the number of pens that person is carrying.
5. Don't be irreplaceable, if you cannot be replaced, you cannot be promoted.
6. Eat one live toad the first thing in the morning and nothing worse will happen to you the rest of the day.
7. Everything can be filed under "miscellaneous."
8. Following the rules will not get the job done.
9. Getting the job done is no excuse for not following the rules.
10. If at first you don't succeed, try again. Then quit. No use being a damn fool about it.
11. If it wasn't for the last minute, nothing would get done.
12. If you are good, you will be assigned all the work. If you are really good, you will get out of it.
13. If you cannot get your work done in the first 24 hours, work nights.
14. Important letters that contain no errors will develop errors in the mail.
15. It doesn't matter what you do, it only matters what you say you've done and what you're going to do.
16. Keep your boss's boss off your boss's back.
17. Never delay the ending of a meeting or the beginning of a cocktail hour.
18. No matter how much you do, you never do enough.
19. People who go to conferences are the ones who shouldn't.
20. The last person that quit or was fired will be held responsible for everything that goes wrong.
21. The more crap you put up with, the more crap you are going to get.
22. There will always be beer cans rolling on the floor of your car when the boss asks for a ride home from the office.
23. To err is human, to forgive is not our policy.
24. When confronted by a difficult problem you can solve it more easily by reducing it to the question, "How would the Lone Ranger handle this?"

25. When the bosses talk about improving productivity, they are never talking about themselves.
26. When you don't know what to do, walk fast and look worried.
27. You are always doing something marginal when the boss drops by your desk.
28. You can go anywhere you want if you look serious and carry a clipboard.

Lawyers

Free Advice

A lawyer went to a dinner party and found herself seated next to a doctor. Eventually conversation turned to the nuisance of being continually approached for free professional advice during social situations.

"I just don't know how to handle this gracefully," she confessed to the doctor. "Do you have any advice?"

"Well," said the doctor, "I doubt this will work for you, but I manage to stop them cold with just one word: UNDRESS."

Gentleman's Club

So the other day, three guys (an accountant, an architect, and a lawyer) went to this "Gentleman's' Club." The accountant wanted to impress the other two, so he pulls out a $10 bill. The "dancer" came over, and the accountant licked the $10 and put it on her butt.

Not to be outdone, the architect pulls out a $50 bill. He calls the girl back over, licks the $50, and puts it on her other cheek.

The lawyer gets out his wallet and calls the girl over. As she comes to a stop before him, he pulls out his ATM card, swipes it down her crack, grabs the 60 bucks, and heads for the door.

Three Wishes

Did you hear about the guy on the beach who found a bottle? He rubbed it and, sure enough, out popped a Genie. "I will grant you three wishes," said the Genie. "But there's a catch." The man was ecstatic. "What catch?" he asked.

The Genie replied, "Every time you make a wish, every lawyer in the world will receive DOUBLE what you asked for." "Well, I can live with that! No problem!" replied the elated man.

"What is your first wish?" asked the Genie. "Well, I've always wanted a Ferrari!" POOF! A Ferrari appeared in front of the man. "NOW, every lawyer in the world has TWO Ferrari's," said the Genie. "Next wish?"

"I'd LOVE a million dollars..." replied the man. POOF! One million dollars appeared at his feet. "NOW, every lawyer in the world has TWO MILLION dollars," said the Genie.

"Well, that's okay, as long as I've got MY million," replied the man.

"What is your final wish?"

The man thought long and hard, and finally said, "Well, you know, I've always wanted to donate a kidney..."

Angel Survey

Two lawyers, Frank and Harry, meet for a drink. Frank says, "You know what happened? An angel was sent down to compile a list of the dishonest lawyers on earth. Six months later he dragged himself back to Heaven, exhausted.

'Believe me,' he told God, 'it'd be easier if I just made note of all of the honest lawyers on earth. In fact, I think I could do that in a weekend.' God said, 'Fine.' Come Monday morning, the angel turned in his list and God said, 'That's terrific. Now I think you should send all the lawyers on this list a note of congratulations.'"

Frank pauses and sips his Scotch. Then he says, "There was a postscript to the angel's note. You know what it was?"

Harry says, "No."

"Aha! So you didn't get one either!"

Shipwrecked

Two lawyers were shipwrecked on a desert island. After several weeks, they saw a beautiful mermaid swim by.

One lawyer says to the other, "Hey, let's screw her."

The other one asks, "Out of what?"

Engine Trouble

An airliner was having engine trouble and the pilot instructed the cabin crew to have the passengers take their seats and prepare for and emergency landing.

A few minutes later, the pilot asked the air hostess if everyone was buckled in and ready. "All ready back here, Captain," came the reply, "except for one lawyer, who is passing out business cards.

Q & A

Q: Why does the law society prohibit sex between lawyers and their clients?
A: To prevent clients from being billed twice for essentially the same service.

Q: What is black and brown and looks good on a lawyer?
A: A Doberman.

Q: What do lawyers and sperm have in common?
A: One in 3,000,000 has a chance of becoming a human being.

Q: What do you call a lawyer who doesn't chase ambulances?

A: Retired.

Be a Lawyer

This guy walks into a bar and tries without success to hit on several women. The bartender, who has been observing his lack of success, tells him "Your problem is that you don't have the right profession to impress these ladies. What you need to do is tell them that you have an upscale job, like a doctor or CPA or lawyer. That's the kind of guy these women are looking for."

The guy takes his advice:... so when a woman asked, "What do you do for a living?" He responded "Oh, well, I'm a lawyer."

"OOhhhh" The guy soon ends up in bed with his newfound lady friend. As their activity intensifies, our hero suddenly bursts out in laughter.

"What, what, I don't understand. What's so funny." she asked. "I was just thinking. Here I've only been a lawyer for 3 hours, and already I'm screwing someone."

Oxymorons

Common Phrases # 1

- Act naturally
- Airline food
- Alone together
- Business ethics
- California culture
- Christian Scientists
- Computer jock
- Definite maybe
- Exact estimate
- Genuine imitation
- Government organization
- Military intelligence
- Passive aggression
- Plastic glasses
- Pretty ugly
- Religious tolerance
- Safe sex
- Sanitary landfill
- Small crowd
- Software documentation
- Taped live
- Terribly pleased

- Advanced BASIC
- Almost exactly
- British fashion
- Butt head
- Childproof
- Clearly misunderstood
- Computer security
- Diet ice cream
- Found missing
- Good grief
- Living dead
- New classic
- Peace force
- Political science
- Rap music
- Resident alien
- Same difference
- Silent scream
- Soft rock
- Synthetic natural gas
- Temporary tax increase
- Working vacation

Common Phrases # 2

- Assistant supervisor
- Business ethics
- Dry lake
- Forward lateral
- Full-time day care
- Highly depressed
- Limited lifetime guarantee
- Mandatory volunteer
- Mutual differences
- Nondairy creamer
- Original copy
- Plastic glass
- Silent alarm
- Standard options
- Uninvited guest

- Authentic reproduction
- Death benefits
- Elevated subway
- Friendly fire
- Genuine veneer
- Holy war
- Live recording
- Mercy killing
- New tradition
- Open secret
- Partial cease-fire
- Resident alien
- Sports sedan
- True replica
- Wireless cable

Driving

How To Identify A Driver's Home Town/State

> ➤ Both hands on steering wheel in a relaxed posture, eyes constantly checking the rear-view mirror to watch for visible emissions from their own or another's car: Colorado
> ➤ Both hands on wheel, eyes shut, both feet on brake, quivering in terror: Ohio, but driving in California.
> ➤ One hand constantly refocusing the rear-view mirror to show different angles of the BIG hair, one hand going between mousse, brush, and rat-tail to keep the helmet hair going, both feet on the accelerator, poodle steering the car, chrome .38 revolver with mother of pearl inlayed handle in the glove compartment: Texas female
> ➤ One hand on Latte, one knee on wheel, cradling cell phone, foot on brake, mind on game: Seattle
> ➤ One hand on steering, yelling obscenities, the other hand waving a gun out the window and firing repeatedly, keeping a careful eye out for landmarks along the way so as to be able to come back and pick up any bullets that didn't hit other motorists so as not to litter: Colorado resident on spotting a car with Texas plates.
> ➤ One hand on wheel, one finger out window: Chicago
> ➤ One hand on wheel, one hand hanging out the window, keeping speed steadily at 70mph, driving down the center of the road unless coming around a blind curve, in which case they are on the left side of the road: Texas country male
> ➤ One hand on wheel, one hand in pants, cradling cell phone, brick on accelerator: California, with gun in lap: L.A.
> ➤ One hand on wheel, one hand on horn: New York
> ➤ One hand on wheel, one hand on hunting rifle, alternating between both feet being on the accelerator and both on the brake, throwing a McDonalds bag out the window: Texas city male
> ➤ One hand on wheel, one hand on newspaper, foot solidly on accelerator: Boston

Hitchhiker Witch

There once was a man who saw a stunning young lady hitch-hiking her way into town. Being ever the gentleman, he offered a ride as far as she wanted to go. As soon as she got into his car the attractive young woman warned him, "I am a witch, and I can cast a spell over you and turn you into anything I want."

The driver must have looked extremely skeptical, because the hitchhiker then said, "I can see that you don't believe me. A little proof may be in order." With that, she leaned over and whispered something in his ear.

Then, sure enough, he turned into a motel.

Recently Seen Bumper Stickers # 1

According to my best recollection, I don't remember.
Anything worth doing is worth overdoing.
Being good at being stupid doesn't count.
Cautious: Breathing may be hazardous to your health.
Cute and interesting are two different things.
Don't let people drive you crazy when you know it's in walking distance.
Every time I think I know where it's at, they move it.
Everybody is somebody else's weirdo.
Everybody looks brave holding a machine gun.
For him to get an idea, it would be a surgical process.
Get forgiveness now - tomorrow you may no longer feel guilty.
Get out of my reality!
Gravity always gets me down.
Had this been an actual emergency, we would have fled in terror and you wouldn't have been notified.
Honk if you like peace and quiet.
I believe in getting in hot water; it keeps you clean.
I don't think I'd be so bored if I didn't have so much to do.
I know my biology; it's your biology I don't know.
I think that I think, therefore I think that I am.
I wouldn't recommend sex, drugs or insanity for everyone, but they've always worked for me.

Recently Seen Bumper Stickers # 2

If I can't fix it, it ain't broken.
If it's not nailed down, it's fair game.
If life's a trip, then where's my ticket?
If we don't know it already, chances are we're not interested in learning it.
If you can't be good, be careful. If you can't be careful, give me a call.
If you can't go first class, charge it.
If you don't care where you are, then you ain't lost.
I'm only a hypochondriac when I'm feeling sick.
I'm serious; it was a joke.
It doesn't matter what temperature a room is; it's always room temperature.
It is your right to be stupid.
It's bad luck to be superstitious.
It's beautiful the way it is; why spoil it by making it legal?
It's not when you get up, but when you get down.

I've forgotten what I'm supposed to do with it.
I've given up trying to escape from reality; they always find me anyway.
Keep grandma off the streets. Legalize bingo.
Life isn't weird; it's the people.
Life's a trip and then you run out of Travelers' Checks.
Never trust a nun with a gun.

Recently Seen Bumper Stickers # 3

Quick! Act as if nothing has happened!
Radioactive cats have 18 half-lives.
Schizophrenia beats being alone.
Sex is not the answer. Sex is the question. Yes is the answer.
Support bacteria, it's the only culture some people have.
The more things change, the more they stay insane.
They are only trying to make me LOOK paranoid.
Today is the tomorrow you worried about yesterday.
Tomorrow will be canceled due to lack of interest.
Well, last year I think it was a Tuesday.
When does summertime come to Minnesota you ask?
You cannot be late until you show up.

Recently Seen Bumper Stickers # 4

A bartender is a pharmacist with a limited inventory.
All generalizations are false.
All men are idiots, and I married their King!
As long as there are tests there will be prayer in public schools.
Assassins do it from behind!
Born Free...Taxed to Death.
Change is inevitable, except from vending machines.
Consciousness: that annoying time between naps.
Conserve toilet paper, use both sides.
Cover me. I'm changing lanes.
Don't blame me, I'm from Uranus.
E. coli Happens.
Few women admit their age; few men act it.
First the engagement ring, then the wedding ring, then the suffering.
Forget the Joneses, I keep up with the Simpsons.
Friends don't let Friends drive naked.
Friends help you move. Real friends help you move bodies.
Give me ambiguity or give me something else.
Happiness is a belt-fed weapon.
He who laughs the last thinks slowest.
I didn't fight my way to the top of the food chain to be a vegetarian.
I get enough exercise just pushing my luck!

I love cats... they taste just like chicken.
I still miss my ex, but my aim is improving!
I took an IQ test and the results were negative.

Recently Seen Bumper Stickers # 5

I want to die in my sleep like my grandfather...not screaming and yelling like the passengers in his car.
If were aren't supposed to eat animals, why are they made of meat?
If you don't like the news go out and make some.
IRS: We've got what it takes to take what you've got.
It's as BAD as you think, and they ARE out to get you!
Jack Kevorkian for White House Physician.
Laugh alone and the world things you're an idiot.
Learn from your parent's mistakes - use birth control.
Montana—-At least our cows are sane.
No Radio - already stolen.
Okay, who stopped the payment on my reality check!
Out of my mind. Be back in five minutes.
Reality is a crutch for people who can't handle drugs.
Rehab is for quitters.
Sex is a misdemeanor...the more I miss it, the meaner I get.
Smile, it's the second best thing you can do with your lips.
Some people are alive only because it's illegal to kill.
Sometimes I wake up grumpy; Other times I let him sleep.
Sorry, I don't date outside my species.
The gene pool could use a little chlorine.
The more people I meet the more I like my dog.
Time is the best teacher; unfortunately it kills of it's students!
Time is what keeps everything from happening at once.
We are Microsoft. Resistance is Futile. You will be Assimilated.
When there's a will I want to be in it.
When you do a good deed get a receipt in case heaven is like the IRS.
Wink, I'll do the rest.
Your kid may be an honor student but you're still and IDIOT!

Recently Seen Bumper Stickers # 6

A bird in the hand is safer than one overhead.
But I'm totally unprepared for everyday life.
I have a new philosophy. I'm only going to dread one day at a time.
I talk to myself because I like dealing with a better class of people.
If Murphy's Law can go wrong, it will.
If today was a fish, I'd throw it back in.
If you drink, don't drive. Don't even putt.
I'm prepared for all emergencies.

It was such a lovely day I thought it was a pity to get up.
It's better to have a horrible ending than to have horrors without end.
It's morally wrong to allow suckers to keep their money.
Laugh at your problems, everyone else does.
Never eat more than you can lift.
Never go to bed mad, stay up and fight.
Never sleep with anyone crazier than yourself.
Stop crime at its source! Support Planned Parenthood.
The only reason people get lost in thought is because it's unfamiliar territory.
The whole purpose of your life is to serve as a warning to others.
There's no point in being grownup if you can't be childish sometimes.
Until you walk a mile in another man's moccasins you can't imagine the smell.
When asked if he had missed school lately, the boy said `Not a bit.`
You're being followed; cut out the hanky-panky for a few days.
You've been leading a dog's life. Stay off the furniture.

Recently Seen Bumper Stickers # 7

All things are possible, except skiing through a revolving door.
Confidence is the feeling you have before you understand the situation.
Death is God's way of telling you not to be such a wise guy.
Don't tell me any big lies today. Small ones can be just as effective.
Everything takes longer than you think.
Excellent time to become a missing person.
Give your child mental blocks for Christmas.
If at first you don't succeed, redefine success.
If you are feeling good, don't worry. You'll get over it.
It is impossible to make anything foolproof because fools are so ingenious.
Just because you're paranoid doesn't mean that they AREN'T after you.
Look out! Behind you!
My opinions may have changed, but not the fact that I am right.
Never hit a man with glasses. Hit him with a baseball bat.
Never put off till tomorrow what you can avoid all together.
Nothing is as easy as it looks.
Of course there is no reason for it, it's just company policy.

Recently Seen Bumper Stickers # 8

The shortest distance between two points is under construction.
There cannot be a crisis next week. My schedule is already full.
They told me I was gullible .. and I believed them.
To err is human, to forgive is not Company Policy.
Today is an excellent day to have a rotten day.
Today is the first day of the rest of the mess.
When in doubt, use brute force.
When you do not know what you are doing, do it neatly.

You know it's a bad day when Suicide Prevention puts you on hold.
You know it's a bad day when the blind date turns out to be your ex-wife.
You know it's a bad day when the sun comes up in the west.
You know it's a bad day when you jump out of bed and miss the floor.
You know it's a bad day when you put both contact lenses in the same eye.
You know it's a bad day when you put your bra on backwards and it fits better.
You know it's a bad day when your income tax refund check bounces.
You know it's a bad day when your pet rock snaps at you.
Your lucky number has been discontinued.

Recently Seen Bumper Stickers # 9

24 hours in a day ... 24 beers in a case ... coincidence?
All those who believe in psycho kinesis raise my hand.
Ambition is a poor excuse for not having enough sense to be lazy.
Beauty is in the eye of the beer holder ...
Black holes are where God divided by zero.
Corduroy pillows: They're making headlines!
Depression is merely anger without enthusiasm
Eagles may soar, but weasels don't get sucked into jet engines
Early bird gets the worm, but the second mouse gets the cheese
Energizer Bunny arrested, charged with battery.
Everyone has a photographic memory. Some just don't have film.
How do you tell when you run out of invisible ink?
I almost had a psychic girlfriend but she left me before we met
I couldn't repair your brakes, so I made your horn louder.
I drive way too fast to worry about cholesterol
I intend to live forever - so far, so good
I love defenseless animals, especially in a good gravy
I poured Spot remover on my dog. Now he's gone.
I tried sniffing Coke once, but the ice cubes got stuck in my nose.
I used to have an open mind but my brains kept falling out.
If Barbie is so popular, why do you have to buy her friends?
If I worked as much as others, I would do as little as they.
If you ain't makin' waves, you ain't kickin' hard enough!
If you choke a smurf, what color does it turn?
I'm not cheap, but I am on special this week
Join the Army, meet interesting people, kill them.
Laughing stock: cattle with a sense of humor.
Many people quit looking for work when they find a job.
Mental backup in progress - Do Not Disturb!
Mind Like A Steel Trap - Rusty And Illegal In 37 States
OK, so what's the speed of dark?
Quantum Mechanics: The dreams stuff is made of.
Shin: a device for finding furniture in the dark.
Support bacteria - they're the only culture some people have

Televangelists: The Pro Wrestlers of religion.
The only substitute for good manners is fast reflexes.
Wear short sleeves! Support your right to bare arms!
What happens if you get scared half to death twice?
When everything's coming your way, you're in the wrong lane.
When I'm not in my right mind, my left mind gets pretty crowded.
Who is General Failure and why is he reading my hard disk?
Why do psychics have to ask you for your name?

Things to Ponder

Observations and Comments # 1

> A neurotic builds castles in the air. A psychotic lives in castles in the air. And a psychiatrist is the guy who collects the rent.

> Advertising: The science of arresting the human intelligence long enough to get money from it.

> Anyone in good enough condition to run three miles a day is in good enough condition not to have to.

> By the time a man can read women like a book he's too old to start a library.

> Did you ever notice when you blow in a dog's face he gets mad at you? But when you take him in a car he sticks his head out the window!

> Don't spend two dollars to dry clean a shirt. Donate it to the Salvation Army instead. They'll clean it and put it on a hanger. Next morning buy it back for seventy-five cents.

> Have you ever noticed... Anybody going slower than you is an idiot, and anyone going faster than you is a maniac?

> If it weren't for electricity we'd all be watching television by candlelight.

> If looks could kill, a lot of people would die with bridge cards in their hands.

> It takes very little to make a woman happy, and more than is contained in Heaven and Earth to keep her that way.

> Pessimists are the world's happiest people. Ninety percent of the time they are right, and the other ten percent they are pleasantly surprised.

> Psychiatrists say that one of four people is mentally ill. Check three friends. If they're okay, you're it.

> Serendipity: looking in a haystack for a needle and finding the farmer's daughter.

> The human brain is a wondrous instrument. It starts working the moment you wake up and doesn't stop until you get to the office.

> The law, in its majestic equality, forbids the rich as well as the poor to sleep under bridges, beg in the streets, and steal bread.

> The race may not always be to the swift nor the victory to the strong, but that's the way to bet.

> The reason most people play golf is to wear clothes they would not be caught dead in otherwise.

> The second day of a diet is always easier than the first. By the second day you're off it.

> Those who beat their swords into plowshares generally end up plowing for those who didn't.
> Why did kamikaze pilots wore helmets.

Hmmmm

> Before they invented drawing boards, what did they go back to?
> Does the Little Mermaid wear an algebra?
> How do I set my laser printer on stun?
> How is it possible to have a civil war?
> If all the world is a stage, where is the audience sitting?
> If love is blind, why is lingerie so popular?
> If most car accidents occur within five miles of home, why doesn't everyone just move 10 miles away?
> If one synchronized swimmer drowns, do the rest have to drown, too?
> If the #2 pencil is the most popular, why is it still #2?
> If the black box flight recorder is never damaged during a plane crash, why isn't the whole airplane made out of the stuff?
> If you ate pasta and antipasta, would you still be hungry?
> If you try to fail, and succeed, which have you done?
> If you're born again, do you have two bellybuttons?
> Why are hemorrhoids called "hemorrhoids" instead of "asteroids"?
> Why is it called tourist season if we can't shoot at them?
> Why is there an expiration date on sour cream?

Observations and Comments # 2

> Do fish get cramps after eating?
> Does the reverse side also have a reverse side?
> How come abbreviated is such a long word?
> How come you press harder on a remote-control when you know the battery is dead?
> How do "Keep off the grass" signs get where they are?
> If it's zero degrees outside today and it's supposed to be twice as cold tomorrow, how cold is it going to be?
> If you got into a taxi and he started driving backwards, would the taxi driver end up owing you money?
> What would a chair look like if your knees bent the other way.
> Why are there 5 syllables in the word "monosyllabic"?
> Why are they called apartments, when they're all stuck together?
> Why do ballet dancers always dance on their toes? Wouldn't it be easier to just hire taller dancers?
> Why do banks charge you a "non-sufficient funds fee" on money they already know you don't have?

> Why do scientists call it "re"search when looking for something new?
> Why do they call it the Department of Interior when they are in charge of everything outdoors?
> Why is a carrot more orange than an orange?
> Why is it that the guy who comes up behind you while you're waiting for an elevator presses the already lit "up" button as though he somehow has magical powers that you didn't when you pressed it the first time?
> Why is the alphabet in that order?
> Why is there only one Monopolies commission?

Daily Affirmations For A Better Life # 1

> All of me is beautiful and valuable, even the ugly, stupid, and disgusting parts.
> As I learn the innermost secrets of the people around me, they reward me in many ways to keep me quiet.
> As I learn to trust the universe, I no longer need to carry a gun.
> As I let go of my feelings of guilt, I can get in touch with my Inner Sociopath.
> Blessed are the flexible, for they can tie themselves into knots.
> Having control over myself is nearly as good as having control over others.
> I am at one with my duality.
> I am grateful that I am not as judgmental as all those censorious, self-righteous people around me.
> I assume full responsibility for my actions, except the ones that are someone else's fault.
> I can change any thought that hurts into a reality that hurts even more.
> I have the power to channel my imagination into ever-soaring levels of suspicion and paranoia.
> I honor my personality flaws, for without them I would have no personality at all.
> I need not suffer in silence while I can still moan, whimper and complain.
> I no longer need to punish, deceive or compromise myself. Unless, of course, I want to stay employed.
> I will strive to live each day as if it were my 40th birthday.
> In some cultures what I do would be considered normal.
> Joan of Arc heard voices too.
> My intuition nearly makes up for my lack of good judgment.
> Only a lack of imagination saves me from immobilizing myself with imaginary fears.

> The first step is to say nice things about myself. The second, to do nice things for myself. The third, to find someone to buy me nice things.
> When someone hurts me, forgiveness is cheaper than a lawsuit. But not nearly as gratifying.

Observations and Comments # 3

> A bartender is just a pharmacist with a limited inventory.
> Ask me about micro waving cats for fun and profit.
> Atheism is a non prophet organization.
> Diplomacy is the art of saying 'Nice doggie!'... till you can find a rock.
> Don't sweat the petty things and don't pet the sweaty things.
> Friends don't let Friends drive Naked.
> Friends help you move. Real friends help you move bodies.
> Guns don't kill people, postal workers do.
> Hang up and drive.
> Help wanted telepath: you know where to apply
> Horn broken, watch for finger.
> I said "no" to drugs, but they just wouldn't listen.
> I went to a bookstore and asked the saleswoman, "Where's the self-help section?" She said if she told me, it would defeat the purpose.
> I.R.S.: We've got what it takes to take what you've got.
> If all those psychics know the winning lottery numbers, why are they all still working?
> If at first you do succeed, try not to look astonished.
> If man evolved from monkeys and apes, why do we still have monkeys and apes?
> If we aren't supposed to eat animals, why are they made of meat?
> I'm just driving this way to piss you off.
> Is boneless chicken considered to be an invertebrate?
> It's a dog eat dog world out there. And they're short on napkins.
> Jesus is coming, everyone look busy.
> Jesus loves you... everyone else thinks you're an asshole.
> Keep honking, I'm reloading.
> Lord save me from your followers.
> Lottery: A tax on people who are bad at math.
> Married people don't live longer than single people. It just seems longer.
> My kid had sex with your honor student.
> Never trust a stockbroker who's married to a travel agent.
> Reality is a crutch for people who can't handle drugs.
> Sex on television can't hurt you... unless you fall off.

➢ The main reason Santa is so jolly is because he knows where all the bad girls live.
➢ Those who live by the sword get shot by those who don't.
➢ Why doesn't Tarzan have a beard?

Daily Affirmations For A Better Life # 2

➢ A good scapegoat is nearly as welcome as a solution to the problem.
➢ Becoming aware of my character defects leads me to the next step — blaming my parents.
➢ Does my quiet self-pity get to me? Yes? Or should I move up to incessant nagging?
➢ False hope is nicer than no hope at all.
➢ I am learning that criticism is not nearly as effective as sabotage.
➢ I am willing to make the mistakes if someone else is willing to learn from them.
➢ I honor and express all facets of my being, regardless of state and local laws.
➢ I will find humor in my everyday life by looking for people I can laugh at.
➢ Just for today, I will not sit in my living room all day watching TV. Instead I will move my TV into the bedroom.
➢ My body is a temple. Do you want to come over for midnight mass?
➢ The complete lack of evidence is the surest sign that the conspiracy is working.
➢ The next time the universe knocks on my door, I will pretend I am not home.
➢ To have a successful relationship I must learn to make it look like I'm giving as much as I'm getting.
➢ To understand all is to fear all.
➢ Today I will gladly share my experience and advice, for there are no sweeter words than "I told you so."
➢ When I dance through life I do the Texas Two-Step.
➢ Who can I blame for my own problems? Give me just a minute...I'll find someone.
➢ Why should I waste my time reliving the past when I can spend it worrying about the future?

Observations and Comments # 4

➢ Do radioactive cats have 18 half-lives?
➢ Does fuzzy logic tickle?

➤ Does 'virgin wool' come from sheep the shepherd hasn't caught yet?
➤ How did a fool and his money GET together?
➤ How do they get a deer to cross at that yellow road sign?
➤ How do you know when it's time to tune your bagpipes?
➤ If corn oil comes from corn, where does baby oil come from?
➤ If it's tourist season, why can't we shoot them?
➤ If there is no God, who pops up the next Kleenex in the box?
➤ If you become a mere shell of your former self, can you hear the ocean?
➤ If you shoot a mime, should you use a silencer?
➤ If you throw a cat out a car window, does it become kitty litter?
➤ If you're a shallow person, when you die, do they bury you in a shallow?
➤ Is it true that cannibals don't eat clowns because they taste funny?
➤ What do dogs call the hottest days of summer?
➤ What do they use to ship styrofoam?
➤ What was the best thing before sliced bread?
➤ What's another word for thesaurus?
➤ When a cow laughs does milk come up its nose?
➤ When you choke a smurf, what color does it turn?
➤ Who do fish drink like?
➤ Why do kamikaze pilots wear helmets?
➤ Why do they call it a TV set when you only get one?
➤ Why do they sterilize the needles for lethal injections?
➤ Why is abbreviation such a long word?

Alternate Ways To Say Someone is Dumb

➤ A few clowns short of a circus.
➤ A few fries short of a Happy Meal.
➤ An experiment in Artificial Stupidity.
➤ As smart as bait.
➤ Body by Fisher, brains by Mattel.
➤ Couldn't pour water out of a boot with instructions on the heel
➤ Doesn't have all his dogs on one leash.
➤ A few feathers short of a whole duck.
➤ A few peas short of a casserole.
➤ An intellect rivaled only by garden tools.
➤ Big like ox, smart like tractor
➤ Chimney's clogged.
➤ Doesn't have all his cornflakes in one box.
➤ Doesn't know much but leads the league in nostril hair.

- Dumber than a box of hair.

- Forgot to pay his brain bill.

- He fell out of the Stupid tree and hit every branch on the way down.
- His antenna doesn't pick up all the channels.
- If he had another brain, it would be lonely.
- No grain in the silo.

- One taco short of a combination plate.
- Several nuts short of a full pouch.
- Slinky's kinked.
- The cheese slid off his cracker.
- Too much yardage between the goal posts.

- Elevator doesn't go all the way to the top floor.
- Has an IQ of 2, but it takes 3 to grunt.
- Her sewing machine's out of thread.

- His belt doesn't go through all the loops.
- Missing a few buttons on his remote control.
- One Fruit Loop shy of a full bowl.
- Receiver is off the hook.

- Skylight leaks a little.

- Surfing in Nebraska.
- The wheel's spinning, but the hamster's dead.
- Warning: Objects in mirror are dumber than they appear.

Joseph J. Zajac III

News

Please Do The Nightly News Without Using Any Of These Words!

- America's sweetheart
- Break through
- Children
- Conservative
- Deadly
- Diva
- Extremist
- Hearts
- Historic
- Massive
- Right wing
- Sexy
- Special
- Terrible tragedy
- Tragedy
- War against (fill in the blank)

- Battle against (fill in the blank)
- Cancer
- Controversy
- Cyber (fill in the blank)
- Disturbing
- Experts
- Global warming
- Hero
- Make a difference
- Poll
- Sex symbol
- So called
- Survey
- Tough
- Victim
- Women

I will give a free copy of my next book to the person who can write a few intelligent sentences using all of these words. Email your submission to: Internethumor@aol.com.

Good, Bad, and Worse News

1. Bad: You can't find your vibrator. Worse: Your daughter "borrowed" it.
2. Bad: You find a porn movie in your son's room. Worse: You're in it.
3. Bad: Your children are sexually active. Worse: With each other.
4. Bad: Your husband's a cross dresser. Worse: He looks better than you.
5. Bad: Your son's involved in Satanism. Worse: As a sacrifice.
6. Bad: Your wife wants a divorce. Worse: She's a lawyer.
7. Bad: Your wife's leaving you. Worse: For another woman.
8. Bad: Your wife's leaving you. Worse: To enter a convent.
9. Bad: Your wife's arrested for soliciting. Worse: She implicates you.
10. Good: The secretary said "yes." Bad: Your wife says "no."

11. Good: The teacher likes your son.

    Bad: Sexually.
    Worse: He's gay.

12. Good: You came home for a quickie.

    Bad: So did the postman.

13. Good: You came home for a quickie.

    Bad: Your wife walks in.

14. Good: You get a three-day weekend.

    Bad: You get the flu on Friday.

15. Good: You get tickets to the theatre.

    Bad: It's performance art.

16. Good: You go to see a strip show.

    Bad: Your daughter's the headliner.

17. Good: Your boyfriend's exercising.

    Bad: So he'll fit in your clothes.

18. Good: Your car conveniently "runs out of gas."

    Bad: For real.

19. Good: Your child's "waiting for Mr. Right."

    Bad: Your son, that is.

20. Good: Your daughter's on the Pill.

    Bad: She's thirteen.

21. Good: Your neighbor exercises in the nude.

    Bad: She weighs 350 pounds.

22. Good: Your son's doing extra credit work.

    Bad: Making a sex ed video.

23. Good: Your uncle leaves you a fortune.

    Bad: It's counterfeit.

24. Good: Your wife bought a porn video.

    Bad: Your daughter's the star.

25. Good: Your wife likes outdoor sex.

    Bad: You live downtown.

147

Seen on the Internet

e-Commentary # 1

- ➢ Being an object of desire and pleasure for men doesn't make me ALL bad...now does it?
- ➢ God didn't put this package together for just one partner!
- ➢ God was really generous and gave men two heads, but as a joke he only gave them a large enough blood supply to operate one at a time.
- ➢ I wanna be like Barbie cause the bitch has everything.
- ➢ It's a dog-eat-dog world and I 'm wearing milk bone panties!!!!
- ➢ Men are like candy, you always want some....But woman are like chocolate, you can't ever get enough.
- ➢ Redheads do what blondes only dream about.
- ➢ Remember, no matter how good he looks or how bad you want him, some gal somewhere is sick of his crap!
- ➢ Sure god created man before woman, but you always make a rough draft before the Final Masterpiece.
- ➢ When God created MEN, she was only joking around!
- ➢ Why get married and make ONE man miserable when I can stay single and make THOUSANDS miserable?
- ➢ Women need men, like a fish needs a bike, or horses need to fly.

e-Commentary # 2

- ➢ A slut will do it with anyone.. A bitch will do it with anyone but you!
- ➢ Beauty fades, but dumb is forever.
- ➢ I am like the Army—I'll make a man be all he can be!
- ➢ I am not a dumb blonde, just smart enough to be blonde.
- ➢ I don't believe in casual sex, so make sure you wear a tie!
- ➢ I never met a man I didn't like.
- ➢ If at first you don't succeed, destroy all evidence that you tried.
- ➢ If it has tires or testicles...your gonna have trouble with it.
- ➢ Men are like blenders...women feel like they should have one, but most of the time, they're not sure why.
- ➢ Please excuse me, but my world is presently upside-down, so if you don't mind, you'll have to speak to my ass.
- ➢ Sticks and stones may break my bones, but whips and chains excite me.
- ➢ The sex was so good even my neighbors had a cigarette.

Epilog

Given the persistence for a more "PC" society (Politically Correct, not Personal Computer) by the Powers That Be, the following parable (and my favorite story) has much merit in how people should look at life:

My Favorite Joke

The boy rode on the donkey and the old man walked. As they went along they passed some people who remarked it was a shame the old man was walking and the boy was riding. The man and boy thought maybe the critics were right, so they changed positions.

Later, they passed some people that remarked, "What a shame, he makes that little boy walk." They then decided they both would walk!

Soon they passed some more people who thought they were stupid to walk when they had a decent donkey to ride. So, they both rode the donkey.

Now they passed some people that shamed them by saying how awful to put such a load on a poor donkey. The boy and man said they were probably right, so they decided to carry the donkey.

As they crossed the bridge, they lost their grip on the animal and he fell into the river and drowned.

The moral of the story?

If you try to please everyone, you might as well kiss your ass good-bye!

What http Really Means

As defined by Webopedia.com: Short for **H**yper**T**ext **T**ransfer **P**rotocol, the underlying protocal by the World Wide Web. HTTP defines how messages are formatted and transmitted, and what actions Web servers and browsers should take in response to various commands. For example, when you enter a URL in your browser, this actually sends an HTTP command to the Web server directing it to fetch and transmit the requested Web page.

Now, don't you feel better knowing what http really means? You can impress all your colleges at the office with this tidbit of knowledge!

Joke Submissions

I had a lot of fun putting this book together and I hope you had as much fun reading e-Humor! as I had putting together this collection of electronic bits and bytes. Looking forward to publishing future volumes as my collection of jokes from the Internet seems never ending and is constantly reducing the amount of valuable disk space on my computer.

If you have a joke that you wish to submit, I would be more than happy to see if it passes muster for future publications. Because of the free nature of the Internet, I cannot give you credit for a joke if it is published nor will you receive any compensation for your submission other than the satisfaction of seeing your joke in print – if it is good enough to be selected! If the joke really makes me laugh, perhaps you will receive a free copy of my next joke book.

You can submit your comments, suggestions on improving the book and most importantly your jokes to: Internethumor@aol.com

Thank you for buying my book!

ABOUT THE AUTHOR

Joe Zajac was born and raised in "sunny" Syracuse, New York. Desiring to get away from the snow and learn the true meaning of "hurry up and wait," Joe enlisted in the Marine Corps to see the world. He has traveled extensively throughout Europe and continues to do so every few years.

Upon graduating with a degree in International Business from Florida Atlantic University, Joe began answering want ads that contained the word "computer" in the job description. His first position was landed because the hiring manager "liked his briefcase." The rest they say is history ...

On his lighter side, Joe's goal is to invent a new computer technology term that serves to confuse everyone and hopefully generate e-millions.

Joe currently resides in Florida, with his loving and lovely wife Pamela and his loyal and very talkative, attack Dachshund, Athena.

He is patiently waiting for an invitation from David Lettermen of the *Late Show* to meet his talkative attack Dachshund, Athena, so she can say "hellrow" to David.

While waiting to attain those goals, Joe plans to continue writing joke books for your enjoyment.